Energy Poverty

Stefan Bouzarovski

Energy Poverty

(Dis)Assembling Europe's Infrastructural Divide

Stefan Bouzarovski
University of Manchester
Manchester, UK

ISBN 978-3-319-88749-4   ISBN 978-3-319-69299-9   (eBook)
https://doi.org/10.1007/978-3-319-69299-9

© The Editor(s) (if applicable) and The Author(s) 2018 This book is an open access publication
Softcover re-print of the Hardcover 1st edition 2018
Open Access This book is licensed under the terms of the Creative Commons Attribution 4.0 International License (http://creativecommons.org/licenses/by/4.0/), which permits use, sharing, adaptation, distribution and reproduction in any medium or format, as long as you give appropriate credit to the original author(s) and the source, provide a link to the Creative Commons license and indicate if changes were made.
The images or other third party material in this book are included in the book's Creative Commons license, unless indicated otherwise in a credit line to the material. If material is not included in the book's Creative Commons license and your intended use is not permitted by statutory regulation or exceeds the permitted use, you will need to obtain permission directly from the copyright holder.
The use of general descriptive names, registered names, trademarks, service marks, etc. in this publication does not imply, even in the absence of a specific statement, that such names are exempt from the relevant protective laws and regulations and therefore free for general use.
The publisher, the authors and the editors are safe to assume that the advice and information in this book are believed to be true and accurate at the date of publication. Neither the publisher nor the authors or the editors give a warranty, express or implied, with respect to the material contained herein or for any errors or omissions that may have been made. The publisher remains neutral with regard to jurisdictional claims in published maps and institutional affiliations.

Cover illustration: Détail de la Tour Eiffel © nemesis2207/Fotolia.co.uk

Printed on acid-free paper

This Palgrave Macmillan imprint is published by Springer Nature
The registered company is Springer International Publishing AG
The registered company address is: Gewerbestrasse 11, 6330 Cham, Switzerland

Preface

This book is the direct result of nearly five years of work on the Energy Vulnerability and Urban Transitions in Europe project (www.urban-energy.org), generously supported by a Starting Grant from the European Research Council—under the European Union's Seventh Framework Programme (FP7/2007–2013/ERC grant agreement number 313478). The project, commonly known under the acronym EVALUATE (Energy Vulnerability and Urban Transitions in Europe project), has sought to transform scientific knowledge and policy action on energy poverty—a form of material deprivation that affects billions of people across the world.

EVALUATE is a multi-sited study, involving extensive research across a variety of cities and countries. Focusing primarily on four Central and Eastern European cities (Budapest, Gdańsk, Prague and Skopje) the project has undertaken a customized survey with 2435 households, supplemented with insights from in-depth household interviews, 'energy diaries' and energy efficiency audits in the homes of approximately 160 households living in the four cities. EVALUATE has entailed 195 expert interviews in a much wider range of sites across the world, as well as an analysis of micro-data from national and European Union surveys of energy poverty. It has led to more than 200 dissemination activities, while laying the basis for the European Energy Poverty Observatory as well as a new European Co-operation for Science and Technology Action on 'European Energy Poverty: Agenda Co-Creation and Knowledge Innovation'.

EVALUATE benefited from a range of parallel knowledge exchange and dissemination events, funded with the support of the UK Energy

Research Centre (in 2013), the Eaga Charitable Trust (in 2016 and 2017) as well as the Journal of Development Studies Conference Fund at the University of Manchester (in 2017). Such activities were further enhanced by related projects in which I was involved as co-investigator, funded by the UK's Economic and Social Research Council (via the 'Urban Transformation in South Africa Through Co-Designing Energy Services Provision' project) as well as Horizon 2020 ('Calculating and Operationalising the Multiple Benefits of Energy Efficiency in Europe'). Overall, this corpus of activities helped create a motivated and vibrant community of experts and practitioners with a global reach.

Also contributing to the intellectual development of this book has been my engagement as a visiting professor at University of Bergen's Department of Geography (at the Centre for Climate and Energy Transformation), as well as my continued role as an External Professor at the Department of Economic Geography at the University of Gdańsk. The latter was one of our partners in the EVALUATE project, alongside Charles University, the Central European University, the University of Skopje as well as the Centre for Environmental Research and Information 'Eko Svest'. Of note is the recognition afforded to EVALUATE by the University of Krakow, which awarded me a habilitation degree in 2017 in relation to, in part, my work on the project. In 2015, I benefited from the intellectual space associated with a visiting fellowship at the Leibniz Institute for Research on Society and Space. Worthy of particular mention is the rich research environment provided by my home institution—the University of Manchester: the Department of Geography, the Collaboratory for Urban Resilience and Energy, and the Manchester Urban Institute.

EVALUATE has engaged a team of outstanding individuals whose contribution to the project has been immeasurable: the core group has consisted of Neil Simcock, Sergio Tirado Herrero, Harriet Thomson, Saska Petrova and Thomas Maltby. Some of the papers that I co-authored with some of them form the basis for part of the material presented in this book. Additional support in the field has been provided by Gerda Jónász, Nevena Smilevska, Jan Frankowski and Roman Matoušek. Special thanks are also due to the Advisory Board of the project, consisting of Michael Bradshaw, Matthias Braubach, Mark Gaterell, Richard Green, Karen Rowlingson, Iwona Sagan, Luděk Sýkora and Diana Ürge-Vorsatz.

Manchester, UK						Stefan Bouzarovski

Contents

1 Energy Poverty Revisited   1

2 Understanding Energy Poverty, Vulnerability and Justice   9

3 Energy Poverty Policies at the EU Level   41

4 The European Energy Divide   75

5 Concluding Thoughts: Embracing and Capturing Complexity   109

Index   115

List of Acronyms

CEER	Council of European Energy Regulators
CEF	Citizens' Energy Forum
ECC	Energy Consumers' Charter
ECE	Eastern and Central Europe
ECHP	European Community Household Panel
EED	Energy Efficiency Directive
EESC	European Economic and Social Committee
EPEE	European Fuel Poverty and Energy Efficiency project
EVALUATE	Energy Vulnerability and Urban Transitions in Europe project
HBS	Household Budget Survey
IEM	Internal Energy Market
LIHC	Low Income High Cost
PSO	Public Service Obligation
SANCO	Directorate General for Health and Consumers
SILC	Statistics on Income and Living Conditions
TEP	Third Energy Package
VCWG	Vulnerable Consumers Working Group

List of Figures

Fig. 2.1	Dimensions influencing the delivery of energy services to the home, and the emergence of domestic energy deprivation. Originally published in Bouzarovski and Petrova (2015)	18
Fig. 2.2	Thermal energy retrofits can have a significant impact on the amelioration of energy poverty—as has been the case in inner-city Prague (photo by Stefan Bouzarovski)	19
Fig. 2.3	District heating is common in the inner city of Skopje (Macedonia)—one of the case study areas of the Energy Vulnerability and Urban Transitions in Europe project (photo by Stefan Bouzarovski)	20
Fig. 3.1	An energy poverty session at European Energy Week, 2017 (photo by Saska Petrova)	66
Fig. 3.2	Participants at an international conference on energy poverty and vulnerability in Manchester, in 2013 (photo by Stefan Bouzarovski)	67
Fig. 4.1	A composite fuel poverty indicator based on the shares of populations in different EU countries facing selected energy poverty-related problems, with the values of the three 'objective' measures divided by 3. Originally published in Bouzarovski (2014)	85
Fig. 4.2	Percentage of people at risk of poverty versus the energy poverty index. Average for EU member states 2003–2013 for both variables. Originally published in Bouzarovski and Tirado Herrero (2017b)	87
Fig. 4.3	Regional variation in the shares of households that experienced energy burdens above 20 per cent in three	

Fig. 4.4	Central European countries. Originally published in Bouzarovski and Tirado Herrero (2017a)	94
	Regional variation in the shares of households that were energy poor according to the LIHC indicator, in three Central European countries. Originally published in Bouzarovski and Tirado Herrero (2017a)	96
Fig. 4.5	Regional variation in the shares of Hungarian and Polish households that experienced inadequately warm or cool homes, mapped against PPP (purchasing power parity)-adjusted GDP per capita figures. Originally published in Bouzarovski and Tirado Herrero (2017a)	97
Fig. 4.6	Shares of households in the LIHC (right-hand x axis) and high energy burden (left-hand x axis) categories in different settlement size categories, Czechia. Originally published in Bouzarovski and Tirado Herrero (2017a)	98

List of Tables

Table 4.1 Correlation matrix: Pearson's r coefficients of linear correlation between SILC energy poverty indicators and index (columns) and the at-risk-of-poverty rate (rows), calculated upon average values of EU-28 Member States for the period 2003–2013. Originally published in Bouzarovski and Tirado Herrero (2017a) 86

Table 4.2 Percentage of Hungarian households who dedicated more than 10 per cent of their energy expenditure to solid fuels in 2005 and 2011, by income deciles. Originally published in Bouzarovski et al. (2016) 91

Table 4.3 Energy poverty indicators for selected housing typologies in Hungary (expressed as shares of households in the relevant category within all households). Above-average values are italicized and shaded. Originally published in Bouzarovski and Tirado Herrero (2017a) 92

Table 4.4 Energy poverty indicators for selected housing typologies in Hungary (expressed as shares of households in the relevant category within all households). Above-average values of the 'category' shares are italicized and shaded. Originally published in Bouzarovski and Tirado Herrero (2017a) 93

Table 4.5 Housing-related indicators of vulnerability to energy poverty. In the table, 'category' refers to the share of households that are considered vulnerable to the given indicator within the specific socio-demographic category; 'total' refers to the cumulative share of households in the sample (i.e. as a proportion of all households) that are considered vulnerable to the given indicator. Above-average values of the 'category' shares are italicized and shaded. Originally published in Bouzarovski and Tirado Herrero (2017a)   100

CHAPTER 1

Energy Poverty Revisited

Abstract This chapter introduces the political and scientific context in which the book is situated. It defines the terms 'energy poverty' and 'infrastructural divide' while discussing the purpose and structure of the book. The book's central aim is the consolidation and development of debates on European and global energy poverty, by exploring the political and infrastructural drivers and implications of the condition across a variety of spatial scales.

Keywords Energy poverty • Energy vulnerability • Fuel poverty Infrastructure • European Union energy policy

INTRODUCTION

Energy poverty occurs when a household is unable to secure a level and quality of domestic energy services—space cooling and heating, cooking, appliances, information technology—sufficient for its social and material needs. This somewhat contested and broad definition lies at the tip of a vast scientific and policy iceberg, involving complex socio-technical relations that extend across the planet. Energy poverty affects millions of people worldwide, even if the causes and consequences vary depending on context. Historically, the existence of this condition in the Global North has

© The Author(s) 2018
S. Bouzarovski, *Energy Poverty*,
https://doi.org/10.1007/978-3-319-69299-9_1

been attributed to contingencies such as low incomes, energy-inefficient homes and high energy prices, while in the Global South, the infrastructural lack of access to more technologically advanced energy carriers has been seen as the main culprit (Bouzarovski & Petrova, 2015). It is estimated that more than one billion people across the world suffer from this condition; and it has received significant prominence thanks to high-profile international drives such as the 'Sustainable Energy for All' initiative, aimed at 'reducing the carbon intensity of energy while making it available to everyone on the planet' so as to contribute to a 'cleaner, just and prosperous world for all' (Sustainable Energy for All, 2017).

The existence of energy poverty in the 'developed' countries of the Global North was traditionally interpreted within a relatively narrow thematic and geographic register: for a long time, public recognition of the problem was limited to the UK and the Republic of Ireland. The last decade has seen as expansion of scientific and policy debates to a much wider range of countries and regions, particularly in Europe but also in North America, Japan, South Korea, Australia and New Zealand (Hilbert & Werner, 2016; Kim, Lee, Ahn, Lim, & Kim, 2016; Liu, Judd, & Santamouris, 2017; Maxim, Mihai, Apostoaie, & Maxim, 2017; Oppenheim, 2016; Scarpellini, Sanz Hernández, Llera-Sastresa, Aranda, & López Rodríguez, 2017; Teller-Elsberg, Sovacool, Smith, & Laine, 2016; Tirado & Jiménez Meneses, 2016; Williams, Wooliscroft, & Lawson, 2015). Of note is the recent establishment of a European Energy Poverty Observatory—a new initiative supported by the European Union (EU), aimed at providing a public hub for the gathering and dissemination of information on the extent and nature of domestic energy deprivation in Europe. The Observatory collects and publishes Europe-wide energy poverty data, while serving as the focal point of an emergent network of policymakers, research scientists, advocacy groups and community activists interested in the issue. It aims to improve the state of the art on energy poverty detection, measurement and reporting by creating a public forum for the exchange of knowledge on the issue (Energy Vulnerability and Urban Transitions, 2017a).

There is an expectation that the Observatory will become a decision-support tool for the significant amount of new EU-wide energy policy and legislation that will be developed in the near future. This process is part of a wider regulatory drive, reflected in the recent inclusion of energy poverty as a distinct thematic area in the Energy Union and Clean Energy Package proposals and the presence of the term in various EU policy documents since 2009. The public recognition of energy poverty has been supported, in part, by concerted advocacy efforts on the part of non-governmental organizations (NGOs), lobby groups and businesses. Recent examples include the formation of a coalition of unions, anti-poverty organizations and environmental NGOs 'committed to join efforts to fight energy poverty and defend the right to renewable energy for all', in part through 'the recognition of affordable energy as a basic human right in EU legislation' (European Anti-Poverty Network, 2017). Also of significance is the European Energy Poverty Task Force, which combines the efforts of a multinational company, think tank and charitable foundation so as to 'improve people's day-to-day lives, while offering concrete solutions to tackle climate change' (Schneider Electric, 2017).

The rising prominence of energy poverty within European policy and science agendas is likely to catalyse a new tide of discussion and deliberation. Ongoing processes of low-carbon transition have significant social justice implications, many of which intersect with key energy poverty concerns. A number of open questions have yet to be resolved, however. It remains unclear how energy poverty relates to wider dynamics of economic and political restructuring. Also, there is uncertainty over the manner in which energy poverty both affects and is reflected in household consumption practices, as well as existing vulnerabilities and forms of deprivation. There is a need to understand how spatial patterns of energy poverty map onto existing inequalities within and among cities, regions and countries, in light of the known economic and infrastructural embeddedness of the condition. Last but not the least, the link between energy poverty and processes of systemic change in the energy sector is poorly theorized in the social science literature—especially when it comes to the manner in which processes of socio-technical change create spatially embedded forms of inequality.

Purpose of the Book

This book aims to both consolidate and advance debates on European and global energy poverty, by exploring the political and infrastructural drivers and implications of the condition across a variety of spatial scales. It stems from a five-year research programme centring on the European Research Council-funded Energy Vulnerability and Urban Transitions in Europe research project (Energy Vulnerability and Urban Transitions, 2017b), which aims to generate a conceptual shift in the mainstream theorization of domestic energy deprivation—away from the conventional focus on poverty, access and energy efficiency, onto more complex and nuanced issues of resilience and precariousness. The book thus highlights the need for a geographical conceptualization of the different ways in which household-level energy deprivation both influences and is contingent upon disparities occurring at a wider range of spatial scales. There is a strong focus on the relationships among energy transformation, institutional change and place-based factors in determining the nature and location of energy-related poverty and vulnerability.

Within its overarching purpose, the book explores how patterns and structures of energy poverty have changed over time, as evidenced by some of the common measures used to describe the condition. In part, this means exploring the makeup of energy-poor demographics across various social and spatial cleavages. I thus touch upon the regional differences that characterize domestic energy deprivation. More broadly, the book argues that energy sector reconfigurations are both reflected in, and shaped by, various domains of social and political organization, especially in terms of creating poverty-relevant outcomes.

Underpinning all of these aims is a wider argument that the inequalities generated by processes of energy system restructuring have a strong geographical component, as they involve spatial and material formations in addition to income deprivation. I contend that the multidimensional nature of energy poverty makes it difficult to capture the phenomenon via a single indicator, thus suggesting that existing measures can only depict individual facets and experiences associated with the predicament. Following Bouzarovski et al. (2017), I argue in favour of unravelling the wider political and spatial implications of energy poverty in contexts where this condition encompasses a wide range of demographic and spatial strata. This starts from the premise that, despite the recognition that indoor environments represent fluid and open spaces that are connected to broader social and ecological systems (Biehler & Simon, 2010), mainstream work on the dynamics of domestic energy deprivation has largely focused on a relatively narrow range of explanatory factors within the home—particularly micro-economic affordability, as well as the thermal efficiency of the dwelling, heating system or appliances (Boardman, 2010). There is a need, therefore, for establishing how energy poverty is embedded in the broader system of infrastructural provision (Coutard, 2002; Marvin, 2012) and institutional change (Harrison & Popke, 2011) while simultaneously affecting both the consumption structure and state policies that characterize energy flows.

The book moves towards a conceptualization of domestic energy deprivation as a systemic issue that cuts across wider material and policy configurations involved in the provision of housing, the regulation of the energy sector and day-to-day political decisions. The chapters that follow, therefore, explore the embeddedness of energy poverty in socio-spatial path dependencies and reform approaches and its influence on the structure of energy demand as a result of household practices. I am also interested in the relationship between domestic energy deprivation, on the one hand, and the conduct of political debates and government decisions, on the other. More broadly, I hint at the infrastructural and political challenges that underpin the emergence of a common European energy policy. Here, my theoretical approach is predicated upon the emergent field of 'energy geographies' (Calvert, 2015; Pasqualetti & Brown, 2014), which highlights the importance of contingencies such as place, territory, path dependency and uneven development in shaping resource flows and consumption practices alike (Bridge, Bouzarovski, Bradshaw, & Eyre, 2013). Research in this burgeoning domain has provided multilayered accounts of global patterns of energy extraction and demand, as well as the spatial implications of socio-technical transitions towards a low-carbon and sustainable future (ibid.). I also utilize ideas from 'assemblage thinking' (Anderson & McFarlane, 2011; DeLanda, 2006) to argue in favour of conceptualizing energy vulnerability in the EU as a heterogeneous mix of material, technical and institutional components with specific territorial ramifications.

Throughout the book, I highlight the importance of demand-side energy services in shaping both the experience and understanding of energy poverty. As part of this, I focus on the broader injustices throughout the energy chain, so as to move beyond one-dimensional analyses solely dedicated to markets or 'the state' as relevant actors (Bouzarovski, Bradshaw, & Wochnik, 2015). The book thus pays central attention to the EU's policy role in regulating energy as a complex multisectoral issue, thus sustaining the functions that it provides for final consumers (ibid.).

STRUCTURE OF THE BOOK

The remainder of the book consists of four chapters. In Chap. 2, I explore the multiple definitional issues surrounding energy poverty, vulnerability, transitions and justice, while laying out the conceptual framework that has informed my approach towards this book and the research leading up to it. Chapter 3 of the book explores the wider political context in which energy poverty is being addressed, scrutinizing the evolution and functioning of relevant policy landscapes at the EU level. In order to unpack the European 'infrastructural divide'—understood as an amalgamation of social and technical relations that is expressed as a set of geographical differences across a variety of scales—Chap. 4 explores how a combination of social, economic and spatial factors has created a landscape of energy vulnerability in Europe. The chapter also includes a review of energy poverty in various national, regional and urban contexts, focusing on a set of Central and East European countries where energy poverty is widespread. The wider implications of household coping practices and practices of fuel switching are also mentioned in this context. The concluding chapter returns to the aims of the book by emphasizing how the rise and persistence of energy poverty are embedded in wider political and spatial relations, as well as the manner in which various policy decisions are helping dismantle inherited and existing socio-technical divisions in Europe and beyond.

REFERENCES

Anderson, B., & McFarlane, C. (2011). Assemblage and geography. *Area, 43*, 124–127.

Biehler, D. D., & Simon, G. L. (2010). The great indoors: Research frontiers on indoor environments as active political-ecological spaces. *Progress in Human Geography, 35*, 172–192.

Boardman, B. (2010). *Fixing fuel poverty: Challenges and solutions*. London: Routledge.

Bouzarovski, S., Bradshaw, M., & Wochnik, A. (2015). Making territory through infrastructure: The governance of natural gas transit in Europe. *Geoforum, 64*, 217–228.

Bouzarovski, S., Herrero, S. T., Petrova, S., Frankowski, J., Matoušek, R., & Maltby, T. (2017). Multiple transformations: Theorizing energy vulnerability as a socio-spatial phenomenon. *Geografiska Annaler: Series B, Human Geography, 99*, 20–41.

Bouzarovski, S., & Petrova, S. (2015). A global perspective on domestic energy deprivation: Overcoming the energy poverty–fuel poverty binary. *Energy Research & Social Science, 10*, 31–40.

Bridge, G., Bouzarovski, S., Bradshaw, M., & Eyre, N. (2013). Geographies of energy transition: Space, place and the low-carbon economy. *Energy Policy, 53*, 331–340.

Calvert, K. (2015). From 'energy geography' to 'energy geographies' Perspectives on a fertile academic borderland. *Progress in Human Geography*, *40*, 105–125.
Coutard, O. (2002). *The governance of large technical systems.* London: Taylor & Francis.
DeLanda, M. (2006). *A new philosophy of society: Assemblage theory and social complexity.* London/New York: Continuum.
Energy Vulnerability and Urban Transitions. (2017a). *EVALUATE team to host new European energy poverty observatory.* http://wp.me/p3gnoe-DL. Retrieved September 1, 2017.
Energy Vulnerability and Urban Transitions. (2017b). *The EVALUATE project.* http://urban-energy.org/evaluate. Retrieved September 1, 2017.
European Anti-Poverty Network. (2017). *Right to energy for all Europeans.* http://www.eapn.eu/wp-content/uploads/2017/06/EAPN-2017-letter-to-MEPs-Right-to-Energy-Coalition-1225.pdf. Retrieved September 1, 2017.
Harrison, C., & Popke, J. (2011). 'Because you got to have heat': The networked assemblage of energy poverty in Eastern North Carolina. *Annals of the Association of American Geographers*, *101*, 1–13.
Hilbert, A., & Werner, M. (2016). Turn up the heat! Contesting energy poverty in Buffalo, NY. *Geoforum*, *74*, 222–232.
Kim, J. S., Lee, I. H., Ahn, Y. H., Lim, S. E., & Kim, S. D. (2016). An analysis of energy consumption to identify urban energy poverty in Seoul. *International Journal of Urban Sciences*, *20*, 129–140.
Liu, E., Judd, B., & Santamouris, M. (2017). Challenges in transitioning to low carbon living for lower income households in Australia. *Advances in Building Energy Research*, *0*, 1–16.
Marvin, S. (2012). Conceptual framework: Governance, transitions and cities. Introduction. In S. Guy, S. Marvin, W. Medd, & T. Moss (Eds.), *Shaping urban infrastructures: Intermediaries and the governance of socio-technical networks* (pp. 15–16). London: Routledge.
Maxim, A., Mihai, C., Apostoaie, C.-M., & Maxim, A. (2017). Energy poverty in Southern and Eastern Europe: Peculiar regional issues. *European Journal of Sustainable Development*, *6*, 247.
Oppenheim, J. (2016). The United States regulatory compact and energy poverty. *Energy Research & Social Science*, *18*, 96–108.
Pasqualetti, M. J., & Brown, M. A. (2014). Ancient discipline, modern concern: Geographers in the field of energy and society. *Energy Research & Social Science*, *1*, 122–133.
Scarpellini, S., Sanz Hernández, M. A., Llera-Sastresa, E., Aranda, J. A., & López Rodríguez, M. E. (2017). The mediating role of social workers in the implementation of regional policies targeting energy poverty. *Energy Policy*, *106*, 367–375.
Schneider Electric. (2017). *Schneider electric actively develops its programme to fight energy poverty.* http://goo.gl/8KzmnA. Retrieved September 1, 2017.

Sustainable Energy for All. (2017). *Our mission.* http://www.se4all.org/our-mission. Retrieved September 1, 2017.

Teller-Elsberg, J., Sovacool, B., Smith, T., & Laine, E. (2016). Fuel poverty, excess winter deaths, and energy costs in Vermont: Burdensome for whom? *Energy Policy, 90,* 81–91.

Tirado, S., & Jiménez Meneses, L. (2016). Energy poverty, crisis and austerity in Spain. *People, Place and Policy, 10,* 42–56.

Williams, J., Wooliscroft, B., & Lawson, R. (2015). Contrasting approaches to fuel poverty in New Zealand. *Energy Policy, 81,* 38–42.

Open Access This chapter is distributed under the terms of the Creative Commons Attribution 4.0 International License (http://creativecommons.org/licenses/by/4.0/), which permits use, duplication, adaptation, distribution and reproduction in any medium or format, as long as you give appropriate credit to the original author(s) and the source, a link is provided to the Creative Commons license and any changes made are indicated.

The images or other third party material in this chapter are included in the work's Creative Commons license, unless indicated otherwise in the credit line; if such material is not included in the work's Creative Commons license and the respective action is not permitted by statutory regulation, users will need to obtain permission from the license holder to duplicate, adapt or reproduce the material.

CHAPTER 2

Understanding Energy Poverty, Vulnerability and Justice

Abstract This chapter outlines past and current definitional issues at the nexus of energy poverty, energy vulnerability, energy justice and energy transitions. It traces the historical development of scientific understandings centring on these topics, while exploring their interactions and interdependencies. The chapter starts from the multiple definitional controversies surrounding fuel poverty and energy poverty, to then discuss the different ways in which notions of energy vulnerability and energy justice have enriched traditional understandings. The latter has been achieved, in part, thanks to a fuller appreciation of the services and production chains via which energy circulates across territories.

Keywords Energy poverty • Energy vulnerability • Energy justice • Energy transitions • Energy services

INTRODUCTION

As I pointed out in Bouzarovski (2014), the nexus between energy and poverty has historically been riddled with conceptual discord. For a long time, politicians and scientists alike failed to recognize that a unique set of issues existed at the intersection of these two domains. A government minister in the UK infamously claimed that 'people do not talk of "clothes poverty" or "food poverty" and I do not think that it is useful to talk of "fuel poverty" either' (Campbell, 1993, p. 58). The establishment of a clear 'fuel poverty' definition in the British academic and decision-making polity (Boardman, 1991) can therefore be considered a pioneering achievement: not only did it necessitate the creation of new state policy, but it also opened the path for scientific debate over the causes, components, symptoms and consequences of domestic energy deprivation that mattered when stipulating what the condition entails.

The official interpretation of fuel poverty in the UK—where this condition is principally seen as the inability to purchase affordable warmth—has proven remarkably resilient despite being challenged in various fora. In the British context, fuel poverty has traditionally been described as a situation in which a household needs to spend more than 10 per cent of its total income (before housing costs) on all fuel used to heat its homes to an acceptable level. Two aspects of this definition are especially significant, not the least in terms of the amount of controversy they have attracted: First, 'needing to spend' refers not to actual expenditure, but to a hypothetical level that is closely related, inter alia, with the thermal energy efficiency of the dwelling. Second, 'acceptable level' is taken to mean that the home is heated in line with the standards recommended by the World Health Organization (WHO)—18 °C for bedrooms and 20–21 °C for living rooms (Boardman, 2010).

The basic principles of this definition were challenged by a government-sponsored review undertaken by John Hills (2012) at the London School of Economics. This extensive investigation, involving multiple stages of consultation with experts and advocacy organizations, concluded that the existing UK definition has made the fuel poverty measure too sensitive to movements in gas and electricity bills as well as 'the precise assumptions made for what are seen as adequate temperatures for people to live at, and the incomes reported to a survey that is mainly not focussed on income measurement' (Hills, 2012, p. 8). It proposed that the government adopt a new 'Low Income High Cost' (LIHC) indicator about the extent of fuel poverty, which would consider households poor if (i) their 'required fuel costs' are above the median level for the entire population; and (ii) spending that amount would leave them 'with a residual income below the official poverty line' (ibid., p. 9). However, the approach attracted a significant amount of controversy, since it led to a significant reduction in the projected number of fuel-poor households, against a background context where the government 'cut overall support reaching the fuel poor in England by 26 per cent and cut the energy efficiency budget reaching fuel poor homes, the most effective long term solution for tackling fuel poverty, by 44 per cent' (Jansz & Guertler, 2012, p. 2). These debates reflect a broader unease in the academic and policy community, concerning the methods and approaches for measuring the extent of energy poverty (Tirado Herrero, 2017; Maxim et al., 2016; Thomson, Snell, & Liddell, 2016).

In addition to the notion of 'fuel poverty'—and as noted above—a raft of similar, but not entirely identical, concepts have been used to describe this condition in other settings, including, inter alia, notions of 'energy precariousness', 'energy precarity', 'energy deprivation'. There also exist more narrow terms that refer to some of its symptoms, such as 'cold homes', 'energy non-payment' or 'energy disconnection' (Bouzarovski, 2014; Petrova, 2017; Wilhite, Shove, Lutzenhiser, & Kempton, 2000). However, one of the most common scientific understandings of 'energy poverty' is one that focuses not on issues of fuel affordability, but rather explores which factors determine the quality and type of energy services received in the home. As we highlighted in Bouzarovski and Petrova (2015), a number of international development organizations and scholars have been focusing on the persistent deficiency of energy infrastructure provision across large parts of Africa, Asia and South America. Despite a long history of international involvement and high-profile political attention, more than 1.2 billion people across the world still lack access to electricity, while a further 2.8 billion have no choice other than traditional biomass for cooking and heating (World Bank, 2014).

Energy poverty in the Global South has received significant academic and policy attention (Gunningham, 2013; Pachauri & Spreng, 2004; Sagar, 2005), often as a result of its extensive impacts on well-being and health: the inability to access modern fuels in the home means that households are often forced to rely on open fires, which in turn leads to high levels of indoor air pollution. Thus, fumes and smoke from open cooking fires are estimated to contribute to the deaths of 1.3 million people per year, predominantly women and children (González-Eguino, 2015). These circumstances are deemed to exert significant impacts on issues such as personal safety, household time budgets, labour productivity and income (Elias & Victor, 2005). As such, energy poverty is a highly gendered problem, with women bearing the brunt of the consequences of inadequate energy access, while suffering from systemic discrimination as well as decreased access to resources and decision-making (Abdullahi, 2017; Clancy, Ummar, Shakya, & Kelkar, 2007; Kumar & Mehta, 2016; Pachauri & Rao, 2013).

Traditionally, energy poverty research in less developed countries has been mainly focused on supply side issues, emphasizing the need for expanding electricity grids based on the experience of developed world countries (Lee, Anas, & Oh, 1999; Munasinghe, 1990; Rahul Sharma & Chan, 2016). Work undertaken by organizations such as the World Bank in particular has highlighted the benefits of extending the coverage of power grids into rural areas (Barnes, 2007; Cook, 2011; Foley, 1992; Pereira, Freitas, & da Silva, 2010), as well as the economic, social and technical barriers to modern energy access (Watson et al., 2011) including the lack of adequate institutional infrastructures and financial capital. This has been demonstrated in case studies from Africa, South America and Southeast Asia alike (Jimenez, 2017; Monyei, Adewumi, Obolo, & Sajou, 2017; Sovacool & Ryan, 2016; Urpelainen, 2016). The principal policies to address energy poverty have been largely driven by the 'electrification for development' imperative, as has been the mainstream identification of the driving forces and consequences of the problem.

In more recent years, scientific and policy attention has turned to the poverty-amelioration potential of micro-generation and renewable energy investment as an alternative to top-down power grid expansion (Adkins, Eapen, Kaluwile, Nair, & Modi, 2010; Bhide & Monroy, 2011). There has been rising awareness of the cultural and political determinants of household energy transitions towards the use of modern fuels in developing countries (Sehjpal, Ramji, Soni, & Kumar, 2014). Also of relevance in this context is scholarship on the distributional and fiscal implications of state-led policies to address energy consumption (Dube, 2003; Karekezi & Kimani, 2002; Lin & Jiang, 2011), as well as the pathways through which increased access to modern fuels contributes to livelihood improvement and human development more generally (Kaygusuz, 2011; Leite et al., 2016; Ouedraogo, 2013; van Els, de Souza Vianna, & Brasil Jr., 2012; Zulu & Richardson, 2013). Debates on the 'other energy crisis' (Eckholm, 1975), therefore, have gradually evolved from a supply-dominated logic underscoring the underdevelopment of technical infrastructures to a more nuanced understanding of the multilayered political economies and relations of power that underpin the emergence and persistence of energy poverty (Sovacool, 2012).

To summarize, global issues of energy equity have been historically considered within two relatively separate scientific and policy registers. While discussions and measures surrounding 'fuel poverty' have been largely seen within the context of unaffordable warmth in the home—and as such have mainly fallen under the remit of economists, sociologists, environmental scientists and engineers—perspectives on energy poverty in the Global South have been closely articulated in relation to the interdisciplinary field of development studies, in addition to focusing on issues of access, equity and investment in socio-technical systems.

The developed–developing world cleavage can be attributed, in part, to specific historical and geographical trajectories in the scientific recognition of domestic energy deprivation. Here, one can find major distinctions regarding the driving forces of energy and fuel poverty, as well as the policies to address them and their impacts on everyday life (where a clear division emerges about the lack of heating vs. the lack of access to electricity). The health consequences of domestic energy deprivation are perhaps the only area in which similarities exist among dominant understandings, even if the energy poverty literature is predominantly preoccupied on indoor air pollution, while fuel poverty is focused on cold air exposure. At the same time, the fuel–energy poverty binary is not universally applicable: in a limited number of cases, the term 'fuel poverty' has been used to capture the policies and measurement approaches that underpin access to non-traditional energy sources (Hailu, 2012), while some authors use 'fuel poverty' and 'energy poverty' interchangeably to describe conditions in either less developed (Nussbaumer, Bazilian, & Modi, 2012; Pachauri & Spreng, 2004) or more developed (Boardman, 2010) countries. Such studies have tended to gloss over—rather than directly engage with—the distinct intellectual and policy traditions that underpin the public recognition and amelioration of the two sets of conditions.

The failure to perceive the complex set of interdependencies between energy and poverty under a common conceptual umbrella has prevented scientists and policymakers from seeing the causes of domestic energy deprivation in an integrated manner. One can thus embrace the emergent terminological messiness developing around energy poverty to argue that the blurring of conventional definitions offers opportunities for advancing scientific and policy debates on the fundamental relationships among energy access, affordability and state policy. This claim is based upon the premise that that all forms of household-scale energy deprivation share the same consequence: a lack of adequate energy services in the home, with its associated discomfort and difficulty. When cross-referenced with the most widely acceptable definition of relative income poverty (a condition with a global definition—see Foster, 1998) fuel and energy poverty alike can be considered under the same conceptual umbrella: as a set of domestic energy circumstances that do not allow for participating in the lifestyles, customs and activities that define membership of society (Buzar, 2007a).

Measuring Energy Poverty: A Challenging Task

In Bouzarovski (2014) I argued that the difficulties associated with defining energy poverty fade in comparison to the complexities involved in measuring its incidence and nature. This has traditionally been an extremely challenging task in light of the specific nature of the problem: it is private (being confined to the domestic domain), temporally and spatially dynamic (by varying over time and in different geographical settings) and culturally sensitive (expectations of energy service are subjective and socially constructed). Nevertheless, three main methods have been used in this context:

- examining the level of energy services in the home (heating, lighting, refrigeration, cooling, etc.) via direct measurement, and comparing the obtained values to a given standard;
- analysing how patterns of household energy expenditure across the population vary in relation to pre-set absolute and relative lines;
- compiling the subjective impressions of households about the level of energy service reached in the home, or collecting self-reported data about housing circumstances that can be used to make indirect judgements about degrees of domestic energy deprivation.

The first approach has not been used on a large scale within the European Union (EU), due to the technical impracticalities and ethical issues associated with it. Adding to this are the difficulties of defining adequate energy service standards, as a result of, in part, cultural specificities: it is known that a home normally considered well-lit and warm in one geographical context may not be seen as such in another (Walker & Day, 2012). However, national statistical agencies across the EU do gather expenditure data via Household Budget Survey (HBS) platforms; combined with census data and information compiled through other research studies, this has allowed experts to identify the social groups and spatial locations suffering from disproportionately high energy costs. Subjective data relevant to energy poverty is also collected by national statistical agencies, as well as Eurostat's Statistics on Income and Living Conditions (SILC) survey, which was preceded between 1994 and 2001 by the European Community

Household Panel (ECHP). The two surveys contain a self-reported indicator about the share of population that is 'unable to keep the home adequately warm' that provides the only directly relevant and internationally comparative tool for judging the extent of energy poverty at the EU scale. Both SILC and ECHP also contain a range of objective data about dwelling quality and the material conditions of households, which means that self-reported views of thermal comfort can be cross-referenced against other built environment and economic strain indicators. However, the quality of these data sets has often been put into doubt by experts working in the field (Herrero, 2017; Thomson, Bouzarovski, & Snell, 2017).

ENERGY SERVICES

If there is one common thread that connects the multiple energy poverty definitions and measurement methods with respect to the underconsumption of energy in the home, it is the pivotal role of 'energy services' (Fell, 2017). Commonly understood as the 'benefits that energy carriers produce for human well being' (Modi, McDade, Lallement, & Saghir, 2005, p. 9), energy services allow for shifting the perspective away from 'fuels' such as 'coal, oil, natural gas, and uranium, and even ... sunlight and wind, along with complex technologies such as hydrogen fuel cells, carbon capture and storage, advanced nuclear reactors, and superconducting transmission lines, to name a few' (Sovacool, 2011a, p. 1659) onto the notion that 'people do not demand energy per se but energy services like mobility, washing, heating, cooking, cooling and lighting' (Haas et al., 2008, p. 4013). As a result, policy goals can start to revolve around issues such as achieving 'adequate levels of light rather than delivering kWh of electricity' (Sovacool, 2011a, p. 1659). This opens the path for approaching the insecurity of demand-side energy services as a distinct societal challenge, allowing for an 'integrated approach to gauge the resilience of a society to meet the needs of its population ... over longer timescales ahead from various interrelated perspectives' (Jansen & Seebregts, 2010, p. 1654).

Energy service approaches also highlight the inadequacy of existing measurement frameworks towards understanding and monitoring energy delivery in the home, which is mainly captured by the number of energy units consumed by the carrier, or the effect that the conversion process has on affected spaces (such as levels of temperature or illumination). Neither of these metrics properly describe the utility and satisfaction received by the final user, partly because the effect of the energy service on his or her requirements—principally a comfortable and well-functioning home—is largely dependent on subjective variables (Karjalainen, 2007; Petrova, Gentile, Mäkinen, & Bouzarovski, 2013). It thus becomes important to consider the individual, household and community-level determinants of energy dynamics in the residential environment, by taking into account environmental, cultural, technical and architectural factors in influencing (Aune, 2007; Lutzenhiser, 1992; Stephenson et al., 2010).

Thinking about energy in terms of the domestic functions that it affords also allows for considering the wider technologies and dynamics involved in the operation of modern homes. The relatively simple (and somewhat out of date) classification of energy services provided by authors such as Reister and Devine (1981) and further enshrined in the 'energy ladder' and 'fuel stacking' models (Masera, Saatkamp, & Kammen, 2000; Nansaior, Patanothai, Rambo, & Simaraks, 2011; Peng, Hisham, & Pan, 2010)—space heating, water heating, space cooling, refrigeration, cooking, drying, lighting, electronic services and appliance services—quickly starts to break down when the relevance of other processes in the home is considered within this context. The inherently multifunctional nature of energy services means that carriers with one primary purpose often serve a range of secondary roles, many of which are not explicitly linked to energy. Thus, a wood-burning stove can provide space heating, hot water, cooking, drying and light, as well as a feeling of cosiness, comfort and a focal point in the home (Cupples, Guyatt, & Pearce, 2007; Petersen, 2008; Reeve, Scott, Hine, & Bhullar, 2013). At the same time, a single energy service can be supplied by a range of different fuels: 'Illumination, for example, can come from candles, kerosene lamps, or electricity' (Sovacool, 2011b, p. 218).

Further testifying to the multifaceted nature of energy services is their complex composition, which entails 'different inputs of energy, technology, human and physical capital, and environment (including natural resources)' (Haas et al., 2008, p. 4013). This means that energy services cannot be understood in solely technological or social terms, but rather represent hybrid 'assemblages' (Bennett, 2005; McFarlane, 2011) operating across a multitude of scales and sites, beyond the confines of the home. As such, they consist of 'composite accomplishments generating and sustaining certain conditions and experiences' (Shove, 2003, p. 165) that are deeply embedded in the 'orchestration of devices, systems, expectations and conventions' (Shove, 2003, p. 165). Hence, energy services embody social practices that are 'configured by the "hanging together" of institutional arrangements, shared cultural meanings and norms, knowledges and skills and varied material technologies and infrastructures' (Walker, 2014, p. 49). The routines that coalesce around systems of provision can thus be studied via a social practice approach that requires 'stepping back from energy itself' (Walker, 2014, p. 49) and moving beyond issues of technological or behavioural efficiency in the series of transformations that lead to the production of useful energy—however important these may be—onto the manner in which end-use energy demand is articulated in time and space (Bridge, Bouzarovski, Bradshaw, & Eyre, 2013; Jalas & Rinkinen, 2013; Walker, 2014).

At a more fundamental level, energy services are driven by needs, which reflect what the recipients of this system of provision effectively require: 'A cooked meal, a well lit room, a fast computer with an internet connection, a cold beer, a warm bed, mechanical power for pumping or grinding' (Sovacool, 2011b, p. 218). As such, the fulfilment of energy needs is a crucial component of the functionings that enable individuals to perform their everyday life and achieve well-being (Nussbaum, 2011; Saith, 2001; Sen, 2009). But needs are themselves closely conditioned by the social practices that inform the social expectations and settings in which energy use takes place. This is particularly obvious in the case of electricity, whose technical versatility and flexibility (Smil, 2003) has often prompted actors on the supply side to actively manage and produce energy demand. Despite its intractability and vastness, therefore, the entire electricity system can be seen 'as an element of electricity-consuming social practices, informing what makes sense for householders to do during (and outside) peak periods' (Strengers, 2012, p. 230).

Energy Vulnerability

Identifying a shared set of energy services required by households in both developed and developing countries can provide an initial step towards the formulation of a planetary approach towards domestic energy deprivation. It is also necessary to highlight any commonalities in the driving forces of energy poverty throughout the supply chain that leads to the delivery of the final service. In developing a common framework for this purpose, Petrova and I (2015) relied on two approaches.

The first is the 'infrastructure and systems of provision' paradigm (Seyfang & Haxeltine, 2012; Southerton, Chappells, & Vliet, 2004; Wilhite et al., 2000) which, put briefly, describes the institutional dynamics and material cultures surrounding the rise of commodity-specific chains that connect production, distribution and consumption activities. By assigning a 'vertical' logic (Fine, 1993) to the circulation of commodities and services, systems of provision approaches affirm the multiple interdependencies and standardizations that allow for the delivery of specific goods and services to the final consumer. In the case of energy, they bring to light the complex network of activities, infrastructures and resources necessary to provide households with energy. It also becomes apparent that the energy chain (Chapman, 1989) extends well into the home, involving multiple conversions from fuel carriers into end-use services. A household's energy needs are at the final point of this system, while driving its emergence (see Fig. 2.1).

Energy vulnerability thinking provides the second lynchpin of our framework. This approach helps draw a distinction between energy or fuel poverty as a descriptor of a state within a certain temporal frame, on the one hand, and vulnerability as a set of conditions leading to such circumstances, on the other (Bouzarovski, 2013; Hall, Hards, & Bulkeley, 2013). One of the departure points for the vulnerability approach is the realization

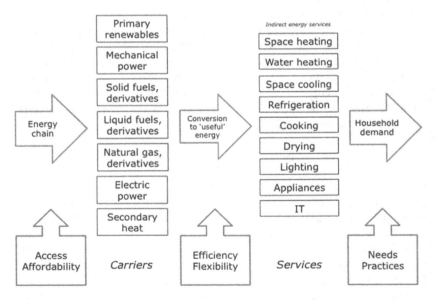

Fig. 2.1 Dimensions influencing the delivery of energy services to the home, and the emergence of domestic energy deprivation. Originally published in Bouzarovski and Petrova (2015)

that households that are described as 'energy service poor' at a given point in time may exit the condition in the future by changing some of their circumstances and vice versa, fuel or energy poverty may affect households that are not described as such at the moment of consideration (Middlemiss & Gillard, 2015). In essence, therefore, energy vulnerability thinking is probabilistic: it highlights the factors that affect the likelihood of becoming poor. When combined with the systems of provision approach, energy vulnerability identifies the role of 'horizontal' factors within different components of the energy chain. These extend beyond the affordability–access binary to encompass the nature and structure of the built environment of the home, as well as the articulation of social practices and energy needs.

In the mainstream literature on 'fuel poverty' in the Global North, the dynamics that underpin the condition are mainly identified within the narrow triad of low household incomes, high energy prices and inadequate levels of energy efficiency (Fig. 2.2). But these are only part of the factors that describe the likelihood of experiencing a socially and materially inadequate level of energy services in the home. The interplay between built

Fig. 2.2 Thermal energy retrofits can have a significant impact on the amelioration of energy poverty—as has been the case in inner-city Prague (photo by Stefan Bouzarovski)

environment flexibility and energy-related social practices means that domestic energy deprivation may arise as a result of a mismatch between the heating or cooling system installed in the dwelling, on the one hand, and the energy service needed by the occupant household, on the other. For example, electric night storage heating is not the most economic option for households that only use the home in the evenings (Milne & Boardman, 2000; Osbaldeston, 1984; Rudge, 2012); and district heating systems that do not have individual controls or thermostats may prove unaffordable for residents who end up 'trapped in the heat' at undesirable times of the day (Tirado Herrero & Urge-Vorsatz, 2012).

In situations where the structural fabric of the building, housing tenure and other legal obstacles do not allow for switching to a more suitable heating system, the household affected by the situation may find itself suffering from inadequate energy services even if it is otherwise able to afford the energy that it consumes, while living in a home that is well insulated (Buzar, 2005, 2007a) (see Fig. 2.3). Moreover, bringing needs into the equation leads, inter alia, to the conclusion that individuals who spend

Fig. 2.3 District heating is common in the inner city of Skopje (Macedonia)— one of the case study areas of the Energy Vulnerability and Urban Transitions in Europe project (photo by Stefan Bouzarovski)

a greater degree of the day at home (such as pensioners or unemployed people) or have specific energy requirements (including disability or the presence of small children) are more likely to suffer from domestic energy deprivation than the rest of the population, as their socio-demographic circumstances mean that such households demand above-average amounts of end-use energy (Buzar, 2007b; Roberts, 2008; Wrapson & Devine-Wright, 2014; Yohanis, 2012). This situation can transpire irrespective of the affordability of energy prices, or the lack of residential energy efficiency.

Vulnerability thinking can also destabilize dominant understandings of the driving forces of this condition in developing countries. A recognition of the need for energy as a socially necessitated phenomenon above basic biological requirements problematizes the idea that minimum standards can provide for adequate individual functionings. Given the multiple sociotechnical trajectories through which any given service can be procured, this suggests that understandings of energy poverty measurement and indicator frameworks via the lens of particular carriers (in contributions such as, for instance, Pachauri, 2011) could enter into a dialogue with work on the entirety of household needs and situations across the world. Of particular importance here are claims that the households primarily desire an energy supply that is reliable, affordable and accessible (Sovacool, 2011a) whereby 'the use and security of energy services is not ingrained but rather conditioned strongly by income and relative wealth within societies' (Sovacool, 2011a). The linear logic of the energy ladder model—which implies that households move towards more technologically sophisticated energy services as their incomes increase and higher levels of national economic development are reached (Masera et al., 2000; Nansaior et al., 2011; Sovacool, 2011a)—is also destabilized by the multiple functions enabled by energy services, ranging from domestic comfort to personal identity. For example, the use of traditional biomass is predicated upon 'active decision making on the part of individual households according to their preferences and broader lifestyle considerations' (Hiemstra-van der Horst & Hovorka, 2008, p. 3342) in developing and developed countries alike.

Alongside issues of access to infrastructure (located at the left side of the energy chain) the affordability of energy is a key underpinning of energy vulnerability. This is because the manner in which state bodies and utilities choose to price energy or support particular groups plays a powerful role in determining whether a household is likely to live in conditions of domestic energy deprivation. Injustices of distribution, procedure and recognition (Walker & Day, 2012) become important factors in driving fuel or energy poverty before even considering issues of income, price or efficiency. Indirect subsidies embedded in the energy tariffs, for example, have a significant impact in determining patterns of deprivation (Freund & Wallich, 1996; Ruggeri Laderchi, Olivier, & Trimble, 2013). Also of relevance in this case are fiscal or pricing measures targeting particular types of fuel; while taxes on diesel and petrol—and even natural gas—are generally less harmful to the poor, it has been demonstrated that placing the tax burden onto electric bills often highly disproportionately affects poor households (Grave et al., 2016). It has been argued that 'schemes that put a price on carbon emissions further upstream ... have an effect not only on downstream energy prices but also on all other goods and services owing to the higher price of the energy used in their production' (Büchs, Bardsley, & Duwe, 2011, p. 291). In some cases, fuel or energy poverty assistance schemes can exacerbate the very condition that they are meant to target by privileging particular groups over others. Regulatory obstacles, information scarcity and socio-cultural factors often prevent socially excluded groups from accessing support (Boardman, 2010).

Moving towards a global understanding of energy vulnerability factors also helps highlight the manner in which the driving forces of deprivation can belong to circumstances that are either internal or external to the household. It becomes apparent that external spheres of action tend to be located at the far ends of the provision system—this also includes the domains of needs and practices. Such thinking is not only useful in identifying groups that may be at risk of falling into energy poverty in the future, but can also help place the combination of social, economic, political and infrastructural factors that have contributed to the position of households that are facing the predicament in the present. This is particularly true in the case of developed world urban households living in transitory housing arrangements—mainly young people, tenants in private rental housing and residents of informal settlements—which are difficult to detect and target via conventional policy frameworks (Bouzarovski, Petrova, Kitching, & Baldwick, 2013; Jencks & Peterson, 2001; Petrova, 2017; Visagie, 2008). In developing country contexts, the framework highlights the crucial importance of ensuring that the technical and financial availability of energy carriers is matched with socially necessitated household needs.

Energy Transitions

Processes of structural change in the energy sector—often called 'energy transitions'—have also been known to increase inequality and deprivation; they are hence of key relevance to understandings of energy poverty (Bouzarovski et al., 2017; Bridge et al., 2013). Even though the term 'energy transitions' implies a shift towards a socially desirable end state, there is no consensus among practitioners or academics as to the exact shape of this future as far as the ongoing process of decarbonization is concerned. While such debates have often taken place under the conceptual umbrella of 'sustainability transitions' (Frantzeskaki, Loorbach, & Meadowcroft, 2012; Lawhon & Murphy, 2012), the multilayered social and technical nature of energy provision means that low-carbon policies inherently involve a complex interplay of political interests, institutional forces and governance practices. The suggestion that the long-term transformation of energy systems will prove 'to be a messy, conflictual, and highly disjointed process' (Meadowcroft, 2009, p. 323) destabilizes the notion that what is at stake is a linear movement towards a predefined environmentally sustainable condition. Moreover, even if a certain set of technical requirements is achieved, there may be no underlying change to the regulatory practices that surround energy use: the same type of infrastructural outcome can be achieved via different policy means, and without altering the basic principles of system organization (Bridge et al., 2013).

The new 'energy paradigm' (Helm, 2005), therefore, opens fundamental questions about the manner in which different political interests and social formations interact with technological change. Some of these dilemmas have included the role of the state in exercising different governing capacities in steering socio-technical transitions (Baker, Newell, & Phillips, 2014), the ability of 'community-based initiatives' to generate innovation (Seyfang & Haxeltine, 2012), the manner in which intermediary organizations assist the implementation of low-carbon strategies (Marvin, 2012) as well as the historical forces involved in shaping deep-seated structural shifts in systems of provision (Smil, 2003).

Historically, energy transitions have been associated with far-reaching shifts in the underpinnings of resource production and distribution, as well as their associated economic and human development patterns. It is claimed that the adoption of low-carbon technology solutions and mitigation strategies brings about multiple benefits in the form of enhanced social welfare and reduced inequalities, and that synergies exist between climate change, poverty alleviation and economic development agendas (Tirado Herrero, 2013; Tirado Herrero & Urge-Vorsatz, 2012). Some of the scholarship in this vein has highlighted the key role of end-use energy services in driving wider socio-technical shifts in society: it has been argued that an improved understanding of energy outputs—rather than the dominant focus on energy outputs—can help explain the relatively slow pace of change in some instances, as well as the emergence of unintended consequences (Grubler, 2012). A distinct body of research has explored the political economies of socio-technical transition (Baker et al., 2014; Meadowcroft, 2009): while acknowledging the pivotal contribution of the 'Multi-Level Perspective' (MLP) in this context, such authors have also highlighted the MLP's shortcomings in terms of the 'assumptions about the nature of state capacity, markets, institutions and infrastructural systems' (Power et al., 2016, p. 12) as well as the manner in which the foregrounding of technology places an emphasis on niche innovations without considering the activities of powerful stakeholders 'whose behaviour cannot be easily shaped by the state' (ibid.). They argue that diverse energy pathways are fundamentally shaped by dynamics of 'power, capacity and autonomy that states have to secure and negotiate' (ibid., p. 11).

Juxtaposing the literatures surveyed above—particularly the suggestions that energy transitions are spatially contingent, imbued with political power and driven by end-use energy demand—suggests that economic and social position of actors and formations implicated in such processes may be deeply affected by structural shifts in energy inputs and outputs alike. This can involve different scales: from nation states whose energy supply may be disrupted to regions that have lost their economic base and consumers who are affected by the decreased availability or increased price of certain fuels (Bouzarovski & Tirado Herrero, 2015; Krishnan, 2016; Smil, 2003). Recent years have also seen a range of contributions focusing on the social vulnerabilities arising from the nexus of climate change mitigation, adaptation and energy policy (Byrne & Portanger, 2016). Some of this work has drilled down to the urban scale, to highlight how the governance of metropolitan systems is enmeshed with perceptions and framings of risk (Rocher, 2016). It can thus be argued that transitions render some actors more socially and economically vulnerable to internal shocks and external pressures, creating new inequalities across time and space.

However, the geographic workings of the energy transitions–vulnerability relationship have received little analytical attention, largely because energy vulnerability itself remains poorly theorized. Recent energy vulnerability scholarship—whose detailed consideration would extend beyond the confines of this chapter—has emphasized the importance of considering the problem through a spatial and temporal framework, while discussing its social construction and the need to consider why and how a given entity may become or be considered vulnerable (Christmann, Ibert, Kilper, & Moss, 2012; Philo, 2012; Waite, Valentine, & Lewis, 2014). Energy vulnerability has been used in a very wide range of contexts, as it can refer to the infrastructural determinants of resource supply and import dependence at a variety of scales, as well as the systemic conditions that allow some entities to become more socially and technically precarious than others (Christie, 2009; Hall et al., 2013; Hiteva, 2013). There is a distinctive literature on household energy vulnerability, understood as a set of circumstances that underpin the risk of falling into fuel and energy poverty. Having applied Spiers' (2000) understanding of 'emic' vulnerability to utility services such as heating and cooling, Middlemiss and Gillard (2015) suggest that energy vulnerability can also be articulated via a bottom-up perspective that characterizes experiences of deprivation.

The material embeddedness of energy vulnerability points to the need for considering the condition though a geographical lens. In the remainder of this book, I consider energy vulnerability as a 'socio-spatial formation' situated at the nexus of political decisions, economic inequalities, organizational practices, on the one hand, and the physical features of place and space, on the other. It should be noted that socio-spatial formations have been theorized extensively in the geography literature, although this body of work has rarely been considered the agency of infrastructural systems. Initial use of the term was motivated by the need for exploring how the political and economic shifts associated with globalization gave rise to specific development patterns and practices of contestation at the urban scale (Amin, 1994). More recent work on the topic has drawn upon regulation theory approaches to emphasize the institutional and political reconfigurations that have underpinned the emergence of entrepreneurial urbanism and the move from 'government to governance' (Mcguirk, 2012). At the same time, assemblage thinking has allowed for socio-spatial formations to be considered as heterogeneous and emergent networks involving the interaction of human and non-human entities via a range of distributed agencies (Anderson & McFarlane, 2011; Dittmer, 2013).

Energy Justice

As Neil Simcock and I (2017) have argued, the application of justice theories and principles to the understanding of energy systems is gaining increasing traction in policy and research circles alike, a movement captured through the emerging concept and frame of 'energy justice' (Jenkins, McCauley, Heffron, Stephan, & Rehner, 2016). Work that connects energy poverty with various concepts of justice has focused predominantly on inequalities between social groups, to the detriment of spatial forms of disadvantage. As noted above, energy justice studies are typically concerned with three fundamental forms of justice: distributive justice, procedural justice and justice as recognition (McCauley, Heffron, Stephan, & Jenkins, 2013). Distributive justice relates to fairness in the distribution of resources; procedural justice to fairness in decision-making process and recognition to the degree of respect given to different socio-cultural identities (Schlosberg, 2007).

In recent years, researchers have contended that the issue of energy poverty is a key dimension of the broader energy justice paradigm (Jenkins et al., 2016). Walker and Day's (2012) pioneering contribution claims that, at its core, energy poverty is 'fundamentally a complex problem of distributive injustice' (p. 69); and suggests that this is underpinned by further injustices in recognition and policymaking procedures. Further studies have built upon this work to unpack the philosophical and moral foundations for considering energy poverty to be a form of injustice (Christman & Russell, 2016; Sovacool, Heffron, McCauley, & Goldthau, 2016).

Alongside such conceptual claims, more grounded work has sought to unveil actual cases of injustice in the incidence and lived experiences of energy poverty. Snell, Bevan, and Thomson (2015) demonstrate that energy poverty disproportionately impacts disabled people in England, and suggest that this form of distributive injustice is driven by the misrecognition of disabled groups. Other studies have revealed how subsidies for low-carbon technologies that are funded through levies on household electricity bills take up a greater proportion of income from the poor compared to those on high incomes (Boardman, 2010; Oppenheim, 2016; Preston, White, Thumim, & Bridgeman, 2013; Stockton & Campbell, 2011), despite low-income groups generally having relatively minor carbon footprints (Jacobson, Milman, & Kammen, 2005) and often benefiting less from decarbonization-related interventions (Oppenheim, 2016; Walker, 2008). Similar claims have been made about the costs of building new nuclear capacity (Garman & Aldridge, 2015). These contributions lend support to a 'whole-systems' approach to energy justice, highlighting the ways that an injustice experienced at the household level (in this case, energy poverty) can be the result of decisions and mechanisms operating elsewhere in the energy system (Jenkins et al., 2016; McCauley et al., 2013).

A number of contributions have begun to explore links between energy deprivation and energy justice—where the emphasis has mainly been on issues of distribution rather than recognition or procedural justice. Throughout this body of work, injustices have predominantly been examined and evaluated in terms of inequalities between socio-demographic and/or socio-economic groups. The justice implications of specifically geographical forms of inequality have rarely been examined. Although a substantial body of literature demonstrates how the occurrence and prevalence of energy poverty is uneven across space (Burholt & Windle, 2006; Healy, 2017; Papada & Kaliampakos, 2016; Thomson & Snell, 2013), such work has principally focused on the drivers or consequences of energy poverty itself, and does not explicitly engage with questions of justice and injustice.

In order to understand how injustices are produced in different geographical contexts, however, it is important to illuminate the manner in which spatially uneven exposure to energy poverty is driven by deeper socio-material inequalities. There is widespread evidence to suggest that the environmental features of a place are crucial in shaping vulnerability to energy poverty. This spatially variegated assemblage of material elements can be described via the more generic notion of 'landscape', so as to highlight the 'heterogeneity of socio-energetic relations and their dynamics' (Castán Broto, Salazar, & Adams, 2014, p. 194; also see Bouzarovski, 2014, for a theorization of 'landscapes of vulnerability'). But even if energy poverty is manifested in particular places, the injustices linked to the environmental factors that produce it extend beyond the spatial and temporal horizons of such locales—expressing a contingency that cannot be easily subsumed within the recognition–procedure–distribution triad. This points to yet another way in which a spatial justice approach illuminates landscapes of material deprivation that add to existing understandings of energy justice.

Climatic conditions are perhaps the most obvious example of an 'environmental' characteristic that can determine household-level vulnerabilities to energy poverty. As climate is underpinned by spatial difference and change over time, some places are thus more likely to face elevated risks. But the impact of climatic differences always occurs in interaction with the characteristics of the built environment including the energy efficiency of homes, heating systems and appliances (Boardman, 2010), the 'flexibility' of heating systems and infrastructures (Buzar, 2007b) and the availability of suitable and cost-effective energy carriers (Bouzarovski & Petrova, 2015). These features are all unevenly distributed across space at a variety of scales and themselves reflect variation in the provision of infrastructural services.

There are also multiple variations between and within nation states. Concerning spatial differences within countries, in Greece, Papada and Kaliampakos (2016) have found that areas in colder climatic zones or higher altitudes are characterized by higher numbers of households paying more than 10 per cent of their income on energy bills (also see Katsoulakos, 2011). Healy and Clinch (2004) have studied rates of energy poverty in Ireland, finding that the shares of households affected by the condition vary geographically between 15 and 18.9 per cent, but with more notable differences in terms of absolute figures—rural areas and Dublin record the greatest number of households living in the condition.

Spatial disparities in household incomes and energy prices contribute to the emergence of geographically uneven injustices. Alongside the national scale, these differences also operate within the grain of cities and regions: local concentrations of low-income households are an important feature of elevated degrees of energy poverty in certain places (Morrison & Shortt, 2008; Walker, Liddell, McKenzie, & Morris, 2013). Moreover, there is also evidence to suggest that low-income households often live in the worst quality housing, partly because they lack the financial means to invest in energy efficiency measures (Boardman, 2010).

CONCLUSION

The variegated understandings of energy services, vulnerability, justice and transitions reviewed above all point to the multiple ways in which household susceptibilities to energy poverty are determined by the material characteristics of residential locations and neighbourhoods. These are highly spatially uneven at a variety of scales (Bouzarovski & Cauvain, 2016) while being embedded in political systems and decision-making structures. The multiple spatially embedded characteristics of the place in which people live—including less known issues such as inflexible heating systems, energy-inefficient buildings and a lack of access to more suitable energy carriers—assemble to create situations of inadequate energy services and high costs (Maxim, Mihai, Apostoaie, & Maxim, 2017).

Like other forms of inequality (Dorling & Ballas, 2008; Walker, 2009), therefore, energy poverty is a deeply geographical and political phenomenon. It is unequally distributed and experienced across different places, and is articulated through complex politics of distribution and recognition. One of the main implications of these arguments is that, in terms of vulnerability to energy poverty, where a person lives seems at least as significant as the socio-economic group that they are part of—yet in much of the current literature and policy discourse inequalities and vulnerability tend to be defined in terms of the latter, rather than in socio-technical, housing or locality terms (Moore, 2012). Thus, spatially uneven patterns of energy poverty are the result of processes and injustices operating throughout the whole energy system, along with economic, material and cultural inequalities acting at various scales. This disrupts the production vs. consumption binary (Jenkins et al., 2016) that has traditionally dominated energy studies, while calling attention to the need for understanding how power interests, relations and processes contribute to the rise of energy-related inequalities.

I have also argued that energy transitions are generators of geographically uneven social, political and environmental displacements. These may increase the vulnerability of particular social groups or places; a contingency that is of special relevance to the global movement towards a low carbon future. The geographies of energy poverty, vulnerability and justice, therefore, embody a distinct temporal dimension. The corollary of this claim is that identifying vulnerable areas also needs to take into account predicted changes in energy prices, forms of infrastructure provision and economic inequality. The two chapters that follow return to the European context via a critical examination of the policy context that allows energy poverty to be being addressed and regulated, while reviewing existing knowledge about the extent and nature of the condition across Europe.

References

Abdullahi, A. A. (2017). An analysis of the role of women in curbing energy poverty in Nigeria. *Journal of Sustainable Development Studies, 10,* 46–50.

Adkins, E., Eapen, S., Kaluwile, F., Nair, G., & Modi, V. (2010). Off-grid energy services for the poor: Introducing LED lighting in the Millennium Villages Project in Malawi. *Energy Policy, 38,* 1087–1097.

Amin, A. (1994). *Post-Fordism: A reader*. Oxford: Blackwell.
Anderson, B., & McFarlane, C. (2011). Assemblage and geography. *Area, 43*, 124–127.
Aune, M. (2007). Energy comes home. *Energy Policy, 35*, 5457–5465.
Baker, L., Newell, P., & Phillips, J. (2014). The political economy of energy transitions: The case of South Africa. *New Political Economy, 19*, 791–818.
Barnes, D. F. (2007). The challenge of rural electrification. In D. F. Barnes (Ed.), *The challenge of rural electrification: Strategies for developing countries* (pp. 1–17). Washington, DC: RFF Press.
Bennett, J. (2005). The agency of assemblages and the North American blackout. *Public Culture, 17*, 445.
Bhide, A., & Monroy, C. R. (2011). Energy poverty: A special focus on energy poverty in India and renewable energy technologies. *Renewable and Sustainable Energy Reviews, 15*, 1057–1066.
Boardman, B. (1991). *Fuel poverty: From cold homes to affordable warmth*. London: Belhaven.
Boardman, B. (2010). *Fixing fuel poverty: Challenges and solutions*. London: Routledge.
Bouzarovski, S. (2013). Energy poverty in the European Union: Landscapes of vulnerability. *Wiley Interdisciplinary Reviews: Energy and Environment*. http://onlinelibrary.wiley.com/doi/10.1002/wene.89/abstract
Bouzarovski, S. (2014). Energy poverty in the European Union: Landscapes of vulnerability. *Wiley Interdisciplinary Reviews: Energy and Environment, 3*, 276–289.
Bouzarovski, S., & Cauvain, J. (2016). Spaces of exception: Governing fuel poverty in England's multiple occupancy housing sector. *Space and Polity, 20*, 310–329.
Bouzarovski, S., Herrero, S. T., Petrova, S., Frankowski, J., Matoušek, R., & Maltby, T. (2017). Multiple transformations: Theorizing energy vulnerability as a socio-spatial phenomenon. *Geografiska Annaler: Series B, Human Geography, 99*, 20–41.
Bouzarovski, S., & Petrova, S. (2015). A global perspective on domestic energy deprivation: Overcoming the energy poverty–fuel poverty binary. *Energy Research & Social Science, 10*, 31–40.
Bouzarovski, S., Petrova, S., Kitching, M., & Baldwick, J. (2013). Precarious domesticities: Energy vulnerability among urban young adults. In *Energy justice in a changing climate: Social equity and low-carbon energy* (pp. 30–45). London: Zed Books.
Bouzarovski, S., & Simcock, N. (2017). Spatializing energy justice. *Energy Policy, 107*, 640–648.
Bouzarovski, S., & Tirado Herrero, S. (2015). The energy divide: Integrating energy transitions, regional inequalities and poverty trends in the European Union. *European Urban and Regional Studies, 24*, 69–86.

Bridge, G., Bouzarovski, S., Bradshaw, M., & Eyre, N. (2013). Geographies of energy transition: Space, place and the low-carbon economy. *Energy Policy, 53*, 331–340.
Büchs, M., Bardsley, N., & Duwe, S. (2011). Who bears the brunt? Distributional effects of climate change mitigation policies. *Critical Social Policy, 31*, 285–307.
Burholt, V., & Windle, G. (2006). Keeping warm? Self-reported housing and home energy efficiency factors impacting on older people heating homes in North Wales. *Energy Policy, 34*, 1198–1208.
Buzar, S. (2005). The institutional trap in the Czech rental sector: Nested circuits of power, space and inequality. *Economic Geography, 82*, 381–405.
Buzar, S. (2007a). *Energy poverty in Eastern Europe: Hidden geographies of deprivation*. Aldershot: Ashgate.
Buzar, S. (2007b). When homes become prisons: The relational spaces of post-socialist energy poverty. *Environment and Planning A, 39*, 1908–1925.
Byrne, J., & Portanger, C. (2016). Climate change, Energy policy and justice: A systematic review. *Analyse & Kritik, 36*, 315–344.
Campbell, R. (1993). Fuel poverty and government response. *Social Policy & Administration, 27*, 58–70.
Castán Broto, V., Salazar, D., & Adams, K. (2014). Communities and urban energy landscapes in Maputo, Mozambique. *People, Place and Policy, 8*, 192–207.
Chapman, J. D. (1989). *Geography and energy: Commercial energy systems and national policy*. Harlow: Longman.
Christie, E. H. (2009). Energy vulnerability and EU-Russia energy relations. *Journal of Contemporary European Research, 5*, 274–292.
Christman, B., & Russell, H. (2016). Readjusting the political thermostat: Fuel poverty and human rights in the UK. *Journal of Human Rights in the Commonwealth, 2*. https://doi.org/10.14296/jhrc.v2i2.2273.
Christmann, G. B., Ibert, O., Kilper, H., & Moss, T. (2012). *Vulnerability and resilience from a socio-spatial perspective: Towards a theoretical framework*. Erkner: Leibniz-Institut für Regionalentwicklung und Strukturplanung eV (IRS).
Clancy, J., Ummar, F., Shakya, I., & Kelkar, G. (2007). Appropriate gender-analysis tools for unpacking the gender-energy-poverty nexus. *Gender and Development, 15*, 241–257.
Cook, P. (2011). Infrastructure, rural electrification and development. *Energy for Sustainable Development, 15*, 304–313.
Cupples, J., Guyatt, V., & Pearce, J. (2007). 'Put on a jacket, you wuss': Cultural identities, home heating, and air pollution in Christchurch, New Zealand. *Environment and Planning A, 39*, 2883–2898.
Dittmer, J. (2013). Geopolitical assemblages and complexity. *Progress in Human Geography, 38*, 385–401.

Dorling, D., & Ballas, D. (2008). Spatial divisions of poverty and wealth. In T. Ridge & S. Wright (Eds.), *Understanding poverty, wealth and inequality: Policies and prospects* (pp. 103–134). Bristol: Policy Press.

Dube, I. (2003). Impact of energy subsidies on energy consumption and supply in Zimbabwe. Do the urban poor really benefit? *Energy Policy, 31,* 1635–1645.

Eckholm, E. (1975). *The other energy crisis: Firewood.* Washington, DC: Worldwatch Institute.

Elias, R. J., & Victor, D. G. (2005). *Energy transitions in developing countries: A review of concepts and literature* (Program on energy and sustainable development, working paper). Stanford University, Stanford.

Fell, M. J. (2017). Energy services: A conceptual review. *Energy Research & Social Science, 27,* 129–140.

Fine, B. (1993). Modernity, urbanism, and modern consumption: A comment. *Environment and Planning D: Society and Space, 11,* 599–601.

Foley, G. (1992). Rural electrification in the developing world. *Energy Policy, 20,* 145–152.

Foster, J. E. (1998). Absolute versus relative poverty. *American Economic Review, 88,* 335–341.

Frantzeskaki, N., Loorbach, D., & Meadowcroft, J. (2012). Governing societal transitions to sustainability. *International Journal of Sustainable Development, 15,* 19–36.

Freund, C. L., & Wallich, C. I. (1996). The welfare effects of raising household energy prices in Poland. *The Energy Journal, 17,* 53–77.

Garman, J., & Aldridge, J. (2015). *When the levy breaks: Energy bills, green levies, and a fairer low-carbon transition.* London: IPPR.

González-Eguino, M. (2015). Energy poverty: An overview. *Renewable and Sustainable Energy Reviews, 47,* 377–385.

Grave, K., Breitschopf, B., Ordonez, J., Wachsmuth, J., Boeve, S., Smith, M., ... Schleich, J. (2016). *Prices and costs of EU energy.* Utrecht: Ecofys Netherlands.

Grubler, A. (2012). Energy transitions research: Insights and cautionary tales. *Energy Policy, 50,* 8–16.

Gunningham, N. (2013). Managing the energy trilemma: The case of Indonesia. *Energy Policy, 54,* 184–193.

Haas, R., Nakicenovic, N., Ajanovic, A., Faber, T., Kranzl, L., Müller, A., & Resch, G. (2008). Towards sustainability of energy systems: A primer on how to apply the concept of energy services to identify necessary trends and policies. *Energy Policy, 36,* 4012–4021.

Hailu, Y. G. (2012). Measuring and monitoring energy access: Decision-support tools for policymakers in Africa. *Energy Policy, 47*(Supplement 1), 56–63.

Hall, S. M., Hards, S., & Bulkeley, H. (2013). New approaches to energy: Equity, justice and vulnerability. Introduction to the special issue. *Local Environment, 18,* 413–421.

Healy, J. D. (2017). *Housing, fuel poverty and health: A pan-European analysis.* Abingdon/New York: Routledge.

Healy, J. D., & Clinch, J. P. (2004). Quantifying the severity of fuel poverty, its relationship with poor housing and reasons for non-investment in energy-saving measures in Ireland. *Energy Policy, 32,* 207–220.

Helm, D. (2005). The assessment: The new energy paradigm. *Oxford Review of Economic Policy, 21,* 1–18.

Hiemstra-van der Horst, G., & Hovorka, A. J. (2008). Reassessing the 'energy ladder': Household energy use in Maun, Botswana. *Energy Policy, 36,* 3333–3344.

Hills, J. (2012). *Getting the measure of fuel poverty: Final report of the fuel poverty review.* London: LSE.

Hiteva, R. P. (2013). Fuel poverty and vulnerability in the EU low-carbon transition: The case of renewable electricity. *Local Environment, 18,* 487–505.

Jacobson, A., Milman, A. D., & Kammen, D. M. (2005). Letting the (energy) Gini out of the bottle: Lorenz curves of cumulative electricity consumption and Gini coefficients as metrics of energy distribution and equity. *Energy Policy, 33,* 1825–1832.

Jalas, M., & Rinkinen, J. (2013). Stacking wood and staying warm: Time, temporality and housework around domestic heating systems. *Journal of Consumer Culture.* https://doi.org/10.1177/1469540513509639.

Jansen, J. C., & Seebregts, A. J. (2010). Long-term energy services security: What is it and how can it be measured and valued? *Energy Policy, 38,* 1654–1664.

Jansz, A., & Guertler, P. (2012). *The impact on the fuel poor of the reduction in fuel poverty budgets in England.* London: Association for the Conservation of Energy.

Jencks, C., & Peterson, P. E. (Eds.). (2001). *The urban underclass.* Washington, DC: Brookings Institution Press.

Jenkins, K., McCauley, D., Heffron, R., Stephan, H., & Rehner, R. (2016). Energy justice: A conceptual review. *Energy Research & Social Science, 11,* 174–182.

Jimenez, R. (2017). Barriers to electrification in Latin America: Income, location, and economic development. *Energy Strategy Reviews, 15,* 9–18.

Karekezi, S., & Kimani, J. (2002). Status of power sector reform in Africa: Impact on the poor. *Energy Policy, 30,* 923–945.

Karjalainen, S. (2007). Gender differences in thermal comfort and use of thermostats in everyday thermal environments. *Building and Environment, 42,* 1594–1603.

Katsoulakos, N. (2011). Combating energy poverty in mountainous areas through energy-saving interventions. *Mountain Research and Development, 31,* 284–292.

Kaygusuz, K. (2011). Energy services and energy poverty for sustainable rural development. *Renewable and Sustainable Energy Reviews, 15,* 936–947.

Krishnan, R. (2016). Energy security through a framework of country risks and vulnerabilities. *Energy Sources, Part B: Economics, Planning, and Policy, 11*, 32–37.

Kumar, P., & Mehta, S. (2016). Poverty, gender, and empowerment in sustained adoption of cleaner cooking systems: Making the case for refined measurement. *Energy Research & Social Science, 19*, 48–52.

Lawhon, M., & Murphy, J. T. (2012). Socio-technical regimes and sustainability transitions: Insights from political ecology. *Progress in Human Geography, 36*, 354–378.

Lee, K. S., Anas, A., & Oh, G.-T. (1999). Costs of infrastructure deficiencies for manufacturing in Nigerian, Indonesian and Thai cities. *Urban Studies, 36*, 2135–2149.

Leite, J. G. D. B., Leal, M. R. L. V., Nogueira, L. A. H., Cortez, L. A. B., Dale, B. E., da Maia, R. C., & Adjorlolo, C. (2016). Sugarcane: A way out of energy poverty. *Biofuels, Bioproducts and Biorefining, 10*, 393–408.

Lin, B., & Jiang, Z. (2011). Estimates of energy subsidies in China and impact of energy subsidy reform. *Energy Economics, 33*, 273–283.

Lutzenhiser, L. (1992). A cultural model of household energy consumption. *Energy, 17*, 47–60.

Marvin, S. (2012). Conceptual framework: Governance, Transitions and Cities. Introduction. In S. Guy, S. Marvin, W. Medd, & T. Moss (Eds.), *Shaping urban infrastructures: Intermediaries and the governance of socio-technical networks* (pp. 15–16). London: Routledge.

Masera, O. R., Saatkamp, B. D., & Kammen, D. M. (2000). From linear fuel switching to multiple cooking strategies: A critique and alternative to the energy ladder model. *World Development, 28*, 2083–2103.

Maxim, A., Mihai, C., Apostoaie, C.-M., & Maxim, A. (2017). Energy poverty in Southern and Eastern Europe: Peculiar regional issues. *European Journal of Sustainable Development, 6*, 247.

Maxim, A., Mihai, C., Apostoaie, C.-M., Popescu, C., Istrate, C., & Bostan, I. (2016). Implications and measurement of energy poverty across the European Union. *Sustainability, 8*, 483.

McCauley, D. A., Heffron, R. J., Stephan, H., & Jenkins, K. (2013). Advancing energy justice: The triumvirate of tenets. *International Energy Law Review, 32*, 107–110.

McFarlane, C. (2011). The city as assemblage: Dwelling and urban space. *Environment and Planning D: Society and Space, 29*, 649–671.

Mcguirk, P. (2012). Geographies of urban politics: Pathways, intersections, interventions. *Geographical Research, 50*, 256–268.

Meadowcroft, J. (2009). What about the politics? Sustainable development, transition management, and long term energy transitions. *Policy Sciences, 42*, 323–340.

Middlemiss, L., & Gillard, R. (2015). Fuel poverty from the bottom-up: Characterising household energy vulnerability through the lived experience of the fuel poor. *Energy Research & Social Science, 6*, 146–154.

Milne, G., & Boardman, B. (2000). Making cold homes warmer: The effect of energy efficiency improvements in low-income homes. A report to the Energy Action Grants Agency Charitable Trust. *Energy Policy, 28*, 411–424.

Modi, V., McDade, S., Lallement, D., & Saghir, J. (2005). *Energy services for the Millennium Development Goals*. Washington, DC: The International Bank for Reconstruction and Development/The World Bank/ESMAP.

Monyei, C. G., Adewumi, A. O., Obolo, M. O., & Sajou, B. (2017). Nigeria's energy poverty: Insights and implications for smart policies and framework towards a smart Nigeria electricity network. *Renewable and Sustainable Energy Reviews*. https://doi.org/10.1016/j.rser.2017.05.237.

Moore, R. (2012). Definitions of fuel poverty: Implications for policy. *Energy Policy, 49*, 19–26.

Morrison, C., & Shortt, N. (2008). Fuel poverty in Scotland: Refining spatial resolution in the Scottish Fuel Poverty Indicator using a GIS-based multiple risk index. *Health & Place, 14*, 702–717.

Munasinghe, M. (1990). Rural electrification in the third world. *Power Engineering Journal, 4*, 189.

Nansaior, A., Patanothai, A., Rambo, A. T., & Simaraks, S. (2011). Climbing the energy ladder or diversifying energy sources? The continuing importance of household use of biomass energy in urbanizing communities in Northeast Thailand. *Biomass and Bioenergy, 35*, 4180–4188.

Nussbaum, M. C. (2011). *Creating capabilities*. Cambridge, MA: Harvard University Press.

Nussbaumer, P., Bazilian, M., & Modi, V. (2012). Measuring energy poverty: Focusing on what matters. *Renewable and Sustainable Energy Reviews, 16*, 231–243.

Oppenheim, J. (2016). The United States regulatory compact and energy poverty. *Energy Research & Social Science, 18*, 96–108.

Osbaldeston, J. (1984). Fuel poverty in UK cities. *Cities, 1*, 366–373.

Ouedraogo, N. S. (2013). Energy consumption and human development: Evidence from a panel cointegration and error correction model. *Energy, 63*, 28–41.

Pachauri, S. (2011). Reaching an international consensus on defining modern energy access. *Current Opinion in Environmental Sustainability, 3*, 235–240.

Pachauri, S., & Rao, N. D. (2013). Gender impacts and determinants of energy poverty: Are we asking the right questions? *Current Opinion in Environmental Sustainability, 5*, 205–215.

Pachauri, S., & Spreng, D. (2004). Energy use and energy access in relation to poverty. *Economic and Political Weekly, 39*, 271–278.

Papada, L., & Kaliampakos, D. (2016). Measuring energy poverty in Greece. *Energy Policy, 94*, 157–165.

Peng, W., Hisham, Z., & Pan, J. (2010). Household level fuel switching in rural Hubei. *Energy for Sustainable Development, 14*, 238–244.

Pereira, M. G., Freitas, M. A. V., & da Silva, N. F. (2010). Rural electrification and energy poverty: Empirical evidences from Brazil. *Renewable and Sustainable Energy Reviews, 14*, 1229–1240.

Petersen, L. K. (2008). Autonomy and proximity in household heating practices: The case of wood-burning stoves. *Journal of Environmental Policy & Planning, 10*, 423–438.

Petrova, S. (2017). Encountering energy precarity: Geographies of fuel poverty among young adults in the UK. *Transactions of the Institute of British Geographers*. http://onlinelibrary.wiley.com/doi/10.1111/tran.12196/abstract.

Petrova, S., Gentile, M., Mäkinen, I. H., & Bouzarovski, S. (2013). Perceptions of thermal comfort and housing quality: Exploring the microgeographies of energy poverty in Stakhanov, Ukraine. *Environment and Planning A, 45*, 1240–1257.

Philo, C. (2012). Security of geography/geography of security. *Transactions of the Institute of British Geographers, 37*, 1–7.

Power, M., Newell, P., Baker, L., Bulkeley, H., Kirshner, J., & Smith, A. (2016). The political economy of energy transitions in Mozambique and South Africa: The role of the Rising Powers. *Energy Research & Social Science, 17*, 10–19.

Preston, I., White, V., Thumim, J., & Bridgeman, T. (2013). *Distribution of carbon emissions in the UK: Implications for domestic energy policy*. York: Joseph Rowntree Foundation.

Rahul Sharma, K., & Chan, G. (2016). Energy poverty: Electrification and wellbeing. *Nature Energy, 1*, 16171.

Reeve, I., Scott, J., Hine, D. W., & Bhullar, N. (2013). 'This is not a burning issue for me': How citizens justify their use of wood heaters in a city with a severe air pollution problem. *Energy Policy, 57*, 204–211.

Reister, D. B., & Devine, W. D., Jr. (1981). Total costs of energy services. *Energy, 6*, 305–315.

Roberts, S. (2008). Demographics, energy and our homes. *Energy Policy, 36*, 4630–4632.

Rocher, L. (2016). Governing metropolitan climate-energy transition: A study of Lyon's strategic planning. *Urban Studies*. https://doi.org/10.1177/0042098015624851.

Rudge, J. (2012). Coal fires, fresh air and the hardy British: A historical view of domestic energy efficiency and thermal comfort in Britain. *Energy Policy, 49*, 6–11.

Ruggeri Laderchi, C., Olivier, A., & Trimble, C. (2013). *Balancing act: Cutting energy subsidies while protecting affordability*. Washington, DC: The World Bank.

Sagar, A. D. (2005). Alleviating energy poverty for the world's poor. *Energy Policy, 33*, 1367–1372.
Saith, R. (2001). *Capabilities: The concept and its operationalisation*. Oxford: Queen Elizabeth House.
Schlosberg, D. (2007). *Defining environmental justice: Theories, movements, and nature*. Oxford: Oxford University Press.
Sehjpal, R., Ramji, A., Soni, A., & Kumar, A. (2014). Going beyond incomes: Dimensions of cooking energy transitions in rural India. *Energy, 68*, 470–477.
Sen, A. (2009). *The idea of justice*. Cambridge: Belknap Press/Harvard University Press.
Seyfang, G., & Haxeltine, A. (2012). Growing grassroots innovations: Exploring the role of community-based initiatives in governing sustainable energy transitions. *Environment and Planning-Part C, 30*, 381–400.
Shove, E. (2003). *Comfort, cleanliness and convenience: The social organization of normality*. Oxford: Berg.
Smil, V. (2003). *Energy at the crossroads: Global perspectives and uncertainties*. Cambridge: MIT Press.
Snell, C., Bevan, M., & Thomson, H. (2015). Justice, fuel poverty and disabled people in England. *Energy Research & Social Science, 10*, 123–132.
Southerton, D., Chappells, H., & van Vliet, B. (Eds.). (2004). *Sustainable consumption: The implications of changing infrastructures of provision*. Cheltenham: Edward Elgar Publishing.
Sovacool, B. K. (2011a). Conceptualizing urban household energy use: Climbing the 'Energy Services Ladder'. *Energy Policy, 39*, 1659–1668.
Sovacool, B. K. (2011b). Security of energy services and uses within urban households. *Current Opinion in Environmental Sustainability, 3*, 218–224.
Sovacool, B. K. (2012). The political economy of energy poverty: A review of key challenges. *Energy for Sustainable Development, 16*, 272–282.
Sovacool, B. K., Heffron, R. J., McCauley, D., & Goldthau, A. (2016). Energy decisions reframed as justice and ethical concerns. *Nature Energy, 1*, 16024.
Sovacool, B. K., & Ryan, S. E. (2016). The geography of energy and education: Leaders, laggards, and lessons for achieving primary and secondary school electrification. *Renewable and Sustainable Energy Reviews, 58*, 107–123.
Spiers, J. (2000). New perspectives on vulnerability using emic and etic approaches. *Journal of Advanced Nursing, 31*, 715–721.
Stephenson, J., Barton, B., Carrington, G., Gnoth, D., Lawson, R., & Thorsnes, P. (2010). Energy cultures: A framework for understanding energy behaviours. *Energy Policy, 38*, 6120–6129.
Stockton, H., & Campbell, R. (2011). *Time to reconsider UK energy and fuel poverty policies?* York: Joseph Rowntree Foundation.
Strengers, Y. (2012). Peak electricity demand and social practice theories: Reframing the role of change agents in the energy sector. *Energy Policy, 44*, 226–234.

Thomson, H., Bouzarovski, S., & Snell, C. (2017). Rethinking the measurement of energy poverty in Europe: A critical analysis of indicators and data. *Indoor and Built Environment*. http://journals.sagepub.com/doi/abs/10.1177/14 20326X17699260.
Thomson, H., & Snell, C. (2013). Quantifying the prevalence of fuel poverty across the European Union. *Energy Policy, 52*, 563–572.
Thomson, H., Snell, C., & Liddell, C. (2016). Fuel poverty in the European Union: A concept in need of definition? *People Place and Policy, 10*, 5–24.
Tirado Herrero, S. (2013). *Fuel poverty alleviation as a co-benefit of climate investments: Evidence from Hungary* (Doctoral thesis). Budapest: Central European University.
Tirado Herrero, S. (2017). Energy poverty indicators: A critical review of methods. *Indoor and Built Environment, 26*, 1018–1031.
Tirado Herrero, S., & Urge-Vorsatz, D. (2012). Trapped in the heat: A postcommunist type of fuel poverty. *Energy Policy, 49*, 60–68.
Urpelainen, J. (2016). Energy poverty and perceptions of solar power in marginalized communities: Survey evidence from Uttar Pradesh, India. *Renewable Energy, 85*, 534–539.
van Els, R. H., de Souza Vianna, J. N., & Brasil, A. C. P., Jr. (2012). The Brazilian experience of rural electrification in the Amazon with decentralized generation—The need to change the paradigm from electrification to development. *Renewable and Sustainable Energy Reviews, 16*, 1450–1461.
Visagie, E. (2008). The supply of clean energy services to the urban and peri-urban poor in South Africa. *Energy for Sustainable Development, 12*, 14–21.
Waite, L., Valentine, G., & Lewis, H. (2014). Multiply vulnerable populations: Mobilising a politics of compassion from the 'capacity to hurt'. *Social & Cultural Geography, 15*, 313–331.
Walker, G. (2008). Decentralised systems and fuel poverty: Are there any links or risks? *Energy Policy, 36*, 4514–4517.
Walker, G. (2009). Beyond distribution and proximity: Exploring the multiple spatialities of environmental justice. *Antipode, 41*, 614–636.
Walker, G. (2014). The dynamics of energy demand: Change, rhythm and synchronicity. *Energy Research & Social Science, 1*, 49–55.
Walker, G., & Day, R. (2012). Fuel poverty as injustice: Integrating distribution, recognition and procedure in the struggle for affordable warmth. *Energy Policy, 49*, 69–75.
Walker, R., Liddell, C., McKenzie, P., & Morris, C. (2013). Evaluating fuel poverty policy in Northern Ireland using a geographic approach. *Energy Policy, 63*, 765–774.
Watson, J., Byrne, R., Morgan Jones, M., Tsang, F., Opazo, J., Fry, C., ... & Castle-Clarke, S. (2011). *What are the major barriers to increased use of modern energy services among the world's poorest people and are interventions to overcome these effective? CEE Review 11-004. Collaboration for Environmental Evidence:*

www.environmentalevidence.org/ SR11004.html. Bangor: Collaboration for Environmental Evidence.

Wilhite, H., Shove, E., Lutzenhiser, L., & Kempton, W. (2000). The legacy of twenty years of energy demand management: We know more about individual behaviour but next to nothing about demand. In *Society, behaviour, and climate change mitigation* (pp. 109–126). Dordrecht: Springer.

World Bank. (2014). *Energy—The facts.* http://web.worldbank.org/WBSITE/EXTERNAL/TOPICS/EXTENERGY2/0,contentMDK:22855502~pagePK:210058~piPK:210062~theSitePK:4114200,00.html

Wrapson, W., & Devine-Wright, P. (2014). 'Domesticating' low carbon thermal technologies: Diversity, multiplicity and variability in older person, off grid households. *Energy Policy, 67*, 807–817.

Yohanis, Y. G. (2012). Domestic energy use and householders' energy behaviour. *Energy Policy, 41*, 654–665.

Zulu, L. C., & Richardson, R. B. (2013). Charcoal, livelihoods, and poverty reduction: Evidence from sub-Saharan Africa. *Energy for Sustainable Development, 17*, 127–137.

Open Access This chapter is distributed under the terms of the Creative Commons Attribution 4.0 International License (http://creativecommons.org/licenses/by/4.0/), which permits use, duplication, adaptation, distribution and reproduction in any medium or format, as long as you give appropriate credit to the original author(s) and the source, a link is provided to the Creative Commons license and any changes made are indicated.

The images or other third party material in this chapter are included in the work's Creative Commons license, unless indicated otherwise in the credit line; if such material is not included in the work's Creative Commons license and the respective action is not permitted by statutory regulation, users will need to obtain permission from the license holder to duplicate, adapt or reproduce the material.

CHAPTER 3

Energy Poverty Policies at the EU Level

Abstract This chapter explores the historical evolution and present content of a common European Union (EU) energy poverty agenda. It identifies the principal institutional and political drivers of this process, as well as the ways in which it has been translated into formal legal and policy documents. Also discussed are the key actors involved in promoting energy poverty-relevant policies, as well as recent trends in the development of a more coherent programme to address the difficulties faced by vulnerable groups. The chapter argues that EU energy poverty policy has been limited by the subsidiarity principle, and as such has been largely shaped by instruments related to the Single Market, even if energy efficiency and social policy-related efforts have also played a role.

Keywords Energy poverty • European Single Market • Policy subsidiarity • Energy transitions • European Union

INTRODUCTION

This chapter provides an overview of how EU policy on energy poverty has emerged and developed over time. It details the content and structure of relevant policy tools and legislation in the context of Europe's energy poverty-related infrastructural divide. In Bouzarovski and Petrova (2015b), we argued that in decision-maker and academic

© The Author(s) 2018
S. Bouzarovski, *Energy Poverty*,
https://doi.org/10.1007/978-3-319-69299-9_3

circles alike, the concept of 'EU energy policy' is generally associated with measures to address residential and industrial consumption practices or transnational security issues. There has been little recognition or knowledge of the fact that the EU is becoming increasingly involved in a new strategic effort situated at the intersection of household fuel use, affordability and residential energy efficiency. Given the significant overlap between notions of energy poverty and energy vulnerability, the chapter uses both terms in its review of relevant policy and literature, conceiving them as policies, events or characteristics that jeopardize the affordability or accessibility of energy at any level of the energy system. In mapping 'EU-led policies' the chapter examines explicit EU official documents and legislation on energy vulnerability and energy poverty, as well as related published papers that reference these concepts or have implications for the EU's approach. National policies are referenced only insofar as they have taken EU policy as a basis or have been altered to reflect EU initiatives.

On the basis of these parameters, the chapter first gives a structural account of the major policy initiatives introduced since 1957, including their formal and informal content, any change over time, the driving actors and stakeholders, the relation to energy policy realities 'on the ground' and, where applicable, their perceived success. To do so, it examines primary and secondary EU law, as well as related research and public statements, and supplements these sources with a review of the academic literature. Data is drawn from systematic searches of the EU law and publications database (EUR-Lex), the EU Community Research and Development Information Service (CORDIS), the website archives of the European Commission's Directorate General (DG) for Energy, the catalogue of the British Library, back issues of three high-ranking academic journals in the field (*Energy Policy, the International Journal of Justice and Sustainability* and *Environment and Planning*) and the Google Scholar search engine. Each of the search results is then methodically reviewed and any relevant sources within it added to the collection of resources.

Traditionally, there has been limited discussion of the systemic processes that lie behind the political acceptance of energy poverty at the European scale, especially in terms of the power actors, interests and relations that have driven the increasing prominence of this issue within EU regulation and debates (Bouzarovski & Petrova, 2015a). The academic literature that deals with EU policy on energy poverty contains little research or commentary on EU-level approaches to combat energy poverty or the impact of EU energy policy frameworks upon social inequality. The few exceptions are Darby's (2012) study on the impact of EU metering policy upon fuel-poor households, Bouzarovski, Petrova and Sarlamanov's (2012) critique of EU energy poverty policies, Bouzarovski and Petrova's (2015b) exploration of EU energy poverty work in the context of EU agenda-setting, Hiteva's (2013) examination of renewables policy and its implications for energy vulnerability as well as Thomson, Snell and Liddell's (2016) discussion of definitional issues surrounding energy poverty in the EU. Also of relevance is Dubois and Meier's (2016) work on how energy poverty patterns in Europe are of relevance for policymaking processes.

International-level institutions and organizations active in energy policy have offered a number of assessments and analyses of the energy vulnerability challenge, taking a global perspective and tending to focus upon access to energy in developing countries and the global dimension of security of supply. Such policy recommendations are not generally specific to any given country or system, and do not comment on the construction of EU policy in this area. Similarly, national-level organizations have tended to analyse national-level policy and prevalence, making reference to the EU only where specific laws and regulations are derived from Brussels, such as the Energy Efficiency Directive (EED). For a long time, EU-led policy research in the energy poverty domain was relatively absent—as reflected in the lack of attention devoted to the problem in work by the European Parliamentary Research Service, its internal think tank and the Commission's Joint Research Centre. As is argued in the text that follows, however, this situation has started to change in recent years.

The chapter is chronologically structured, mapping the evolution of energy poverty policy over time, but drawing out key thematic issues and exploring these in more depth when relevant to the key aims of the book. After walking through the foundations of the modern policy framework, the organization of the argument reflects the three main 'sources' of current energy poverty policy, identified as the Third Energy Package (TEP), the EED and various social policy provisions (European Commission, 2013c). A final section provides an analysis of the governance structure that now characterizes energy poverty policy, before short conclusions are offered. Firstly, however, I examine the broader context of EU energy policy and law, which frames energy poverty policy.

The Context: EU Energy Policy

As we pointed out in Bouzarovski and Petrova (2015b), the lack of attention towards EU energy poverty policy has transpired despite the fact that the constituent dynamics of some of the political developments and institutional structures associated with adjacent programmatic sectors are well known. There is, thus, a sizeable body of research on the underlying principles and implementation challenges associated with EU policy in the environmental policy and security domains (McCormick, 2014).

A common EU energy policy did not exist in any coherent form until 2007. After initial integration under the founding treaties, the EU energy policy became marginalized and irrelevant as the dominant energy mix shifted from coal- to oil-based in the early 1960s (Duffield & Birchfield, 2011). While the EU had a solid legal mandate concerning coal, it did not enjoy similar powers in relation to oil. Since Member States were not willing to update the relevant institutional mandates, the EU's role became limited—every decade from the 1950s to the 1990s saw attempts to create a common energy policy but to little avail (Duffield & Birchfield, 2011). Proposals on an energy chapter in the Maastricht Treaty were dropped and a 1995 White Paper was met with indifference and opposition from Member States (Commission of the European Communities, 2005).

The First Energy Package (consisting of directives 96/92/EC and 98/30/EC) was launched in 1997 and sought to harness the momentum of the Single Market and signalled the beginning of a renewed effort in the creation of a common energy policy, as well as a series of developments in energy–environmental and energy–external relations policy. Statements of coherent policy objectives began to be published in the mid-2000s, reaching a peak with the Second Energy Package (consisting of Directive 2003/54/EC, Directive 2003/55/EC, Regulation 1228/2003 and Regulation 1775/2005) in July 2007, and were supported by comparable efforts in the Council and the European Parliament (Duffield & Birchfield, 2011). The Energy Policy Communication was followed by commitment to an action plan by Member States and also formed the basis for the 2008 Energy and Climate Package. Finally, this proliferation of energy instruments and the imminent publication of the TEP (consisting of Directive 2009/72/EC, Directive 2009/73/EC, Regulation 713/2009, Regulation 714/2009 and Regulation 715/2009) were captured by the inclusion of a dedicated title in the Lisbon Treaty in 2009.

Energy was at the centre of the founding treaties—the European Coal and Steel Community (ECSC) and Atomic Energy (Euratom) treaties formed a common policy based on supranational powers and a central authority. However, subsequent accords of this kind did not develop an overarching legal basis for the EU energy policy and when the ECSC Treaty expired in 2002, the Euratom Treaty, relating only to the nuclear energy sector, was the only remaining legal basis (Andoura, Hancher, & van der Woude, 2010). Seeking to address this absence of mandate, the Lisbon Treaty contains for the first time a dedicated Title on energy which sets out the four main aims of the EU's policy. However, it does little to change the parameters of energy policy development, instead constituting 'a carefully crafted compromise between national sovereignty over natural resources and energy taxation on the one hand, and shared EU competence for other areas on the other' (Andoura et al., 2010, p. 19).

The current provisions, embodied in Title XXI (Article 194) of the Treaty on the Functioning of the EU (TFEU), state that (Official Journal of the European Union, 2012):

1. In the context of the establishment and functioning of the internal market and with regard for the need to preserve and improve the environment, Union policy on energy shall aim, in a spirit of solidarity between Member States, to:

a) ensure the functioning of the energy market;
b) ensure security of energy supply in the Union;
c) promote energy efficiency and energy saving and the development of new and renewable forms of energy; and
d) promote the interconnection of energy networks

2. Without prejudice to the application of other provisions in the Treaties, the European Parliament and the Council, acting in accordance with the ordinary legislative procedure, shall establish the measures necessary to achieve the objectives in paragraph 1. Such measures shall be adopted after consultation of the Economic and Social Committee and the Committee or the Regions.

Such measures shall not affect a Member State's right to determine the conditions for exploiting its energy resources, its choice between different energy sources and the general structure of its energy supply.

Additionally, Declaration 35, annexed to the TFEU by the intergovernmental conference that adopted the Lisbon Treaty, states that:

The Conference believes that Article 194 does not affect the right of the Member States to take the necessary measures to ensure their energy supply under the conditions provided for in Article 347.

Article 347 of the TFEU commits Member States to consult with each other so as to protect the functioning of the internal market in the event that a member state has to take action relating to internal disturbances, international tension or the threat of war or the maintenance of law and order. Finally, Article 122 expands upon the notion of solidarity in this context, stating that:

> Without prejudice to any other procedures provided for in the Treaties, the Council, on a proposal from the Commission, may decide, in a spirit of solidarity between Member States, upon the measures appropriate to the economic situation, in particular if severe difficulties arise in the supply of certain products, notably in the area of energy.

As pointed out by Duffield and Birchfield (2011, p. 4), 'the crucial implication of Title XXI is that it removes any equivocation about the legal basis of EU activity in the energy field'. The fact that energy is still a shared competence allows EU energy initiatives to move beyond a 'piecemeal' approach 'based on tenuous links to existing mandates in market liberalization and environmental policy' (ibid.). This overall set of circumstances allowed the European Commission to publish its Communication on the Energy Union in February 2015 (European Commission, 2015). The document calls for a fundamental transformation of Europe's energy system, leading to a global voice on energy on the behalf of the EU, as well as the building of a sustainable and low-carbon economy within the EU. The liberalization of cross-border energy trade is another key tenet of the Communication, as well as the optimization of resource use and the empowerment of citizens in driving energy transitions forward. The reduction of energy bills via new technologies, active market participation, energy efficiency investment and the protection of vulnerable consumers also feature prominently within this document.

Two other overarching frameworks that have guided EU energy policy are the Europe 20-20-20 Strategy (and the Climate and Energy Package adopted within it) and the Energy Roadmap 2050. The 20-20-20 Strategy seeks a 20 per cent reduction in EU greenhouse gas emissions from 1990 levels, an increase in the proportion of EU energy produced from renewable resources to 20 per cent and a 20 per cent improvement in the EU's energy efficiency. The Energy Roadmap also aims to reduce EU emissions by 80 per cent by 2050 via a programme of decarbonization. In a communication produced by the European Commission in 2011, it was stated that 'vulnerable consumers are best protected from energy poverty through a full implementation by Member States of the existing EU energy legislation and use of innovative energy efficiency solutions' while emphasizing that 'the social aspects of energy pricing should be reflected in the energy policy of Member States' since 'energy poverty is one of the sources of poverty in Europe' (European Commission, 2011). Subsequently, the European Parliament:

> welcomes the inclusion of the social dimension in the Energy Roadmap 2050; considers that, in this respect, special attention should be given to energy poverty and employment; insists, with regard to energy poverty, that energy should be affordable for all, and calls on the Commission and the Member States, and on local authorities and competent social bodies, to work together on tailored solutions to counter such issues as electricity and heat poverty, with a special emphasis on low-income, vulnerable households that are most affected by higher energy prices. (European Parliament, 2013a)

ENERGY AS A HUMAN RIGHT

The European Commission's Group on Ethics in Science and New Technologies (EGE) outlined in its 2013 opinion the international human rights framework and the way in which it supports the notion of access to energy as a fundamental human right. While this statement is not made explicitly in the major international treaties, the EGE finds it to be implicit in provisions, such as those in the 1966 Covenants on economic, social and cultural rights, which set out a number of rights essential for the realization of the right to an adequate standard of living, and the Charter of Fundamental Rights of the EU, which states that the Union recognizes and respects access to services of general economic interest (European Group on Ethics in Science and New Technologies to the European Commission, 2013). Though the latter term is not defined, subsequent interpretation by the Commission and the Court has found it to encompass energy provision. The status of the Charter is such that this is not a directly conferred European right, but that the EU may not take legislative steps which curtail access to energy, thus presenting a line of defence for advocates of energy access as a human right (European Group on Ethics in Science and New Technologies to the European Commission, 2013). The EGE also recognizes access to energy and the combatting of energy poverty as one of the key ethical challenges facing EU energy policy.

More recently, a group of advocacy groups led by the European Anti-Poverty Network and Greenpeace have argued that the eradication of energy poverty in Europe would hinge upon implementing 'a right to energy for all citizens' in all EU legislation. They have taken this to mean prohibiting disconnections, maintaining regulated prices for domestic consumers and allowing low-income households to benefit from specific social tariffs. They have also argued that overcoming energy poverty

> will require a holistic political approach, linking social and environmental policy. The economic crisis, ensuing austerity policies and the growing precariousness of the labour market remain the main drivers of the rise of poverty in Europe, however, it is crucial to recognize the role of energy policies in the rise of this issue. ... Awareness is also increasing at national levels, as more and more public bodies, organisations and social movements deepen their understanding of the specificity of energy poverty. But awareness alone will not deliver: it is time for political action to fight energy poverty at the European level. (European Anti-Poverty Network, 2017)

However, the right to energy remains a widely debated issue in the academic literature, since the entity to which the right would be ascribed remains unclear: inter alia, the right to 'energy' itself can consist of legal supplier obligations, end-use services or the possibility of demand (Walker, 2015). Analogous debates of the 'right to water' have also warned against the possibility of introducing a neoliberal lexicon to the debate (Bakker, 2007).

FOUNDATIONS OF EU ENERGY POVERTY POLICY, 1957–2007

Reflecting the broader evolution of energy policy, energy poverty policy as a coherent EU issue did not exist prior to 2007. A EUR-Lex search finds just six documents of secondary EU law mentioning 'energy vulnerability' or 'energy poverty' in the 50 years from 1957; these terms do not feature at all in the primary law of the period.

Neither the First Energy Package nor the Second Energy Package contained any reference to energy vulnerability or energy poverty. Early recognition of this phenomenon in the EU came in the form of a series of policy documents on sustainable development and EU external assistance. The EU Energy Initiative for Poverty Reduction and Sustainable Development, launched in 2002 at the World Summit on Sustainable Development in Johannesburg, linked access to energy in developing countries to achievement of the Millennium Development. The link between poverty reduction and energy is further developed in EU cooperation with countries in the Global South, and later in the Global Energy Efficiency and Renewable Energy Fund.

It was not until 2006 that energy poverty was introduced as an EU issue by the European Commission. In a Communication on prospects for the internal markets in electricity and gas, the Commission pledged to review national approaches to energy poverty and to launch a 'major information and awareness raising campaign' in the run-up to the full market opening in 2007, including the creation of an Energy Consumers' Charter (ECC). A report on economic and social cohesion identified energy poverty as a social issue and these themes were built into the Commission's Communication on an energy policy for Europe.

In its 2007 Communication, 'An energy policy for Europe', the Commission identified sustainability, security of supply and competitiveness as the three main challenges facing EU energy policy (Commission of the European Communities, 2007). In proposing an ECC, the Commission prioritized an understanding of energy as a public service that forms a central part of the modern energy policy framework. The European Council was quick to offer its support to the Commission's latest attempt to kick-start a common policy, agreeing on an action plan for the next two years (Council of the European Union, 2007). The measures proposed were comprehensive, but did not include reference to energy poverty or vulnerable consumers. Nonetheless, the collective momentum provided by the Commission and the Council fed into the establishment of a dedicated title in the Lisbon Treaty, providing a basis for moves towards a coherent common energy policy.

A helpful contribution towards greater public recognition of energy poverty was made by the European Fuel Poverty and Energy Efficiency project, which received European Commission funding under the Intelligent Energy for Europe programme between 2006 and 2009, and focused upon fuel poverty as an inability to adequately heat the home at an affordable cost (Intelligent Energy Europe, 2017). A partnership between agencies in the UK, France, Belgium, Italy and Spain sought to raise the profile of fuel poverty and examine the potential for a common European response. Its final report listed a range of causes and indicators and estimated that fuel poverty affects between 50 and 125 million people in Europe. The project's recommendations identified four core actions that should be undertaken at the EU level—a common definition, a legislative framework, a consistent diagnosis and a fuel poverty special interest group.

As I and Petrova (2015a) have previously pointed out, the European energy poverty sphere has also been influenced by the activities of the Council of Europe (CoE) in the domain of housing policy. This is reflected in a CoE report on the issue (Council of Europe, 2008) drafted by the group of specialists on housing policies for social cohesion. The report aims to aid the improvement of housing access among vulnerable social groups in Europe and has resulted in a set of specific guidelines. These stipulate the key prerequisites for an effective housing policy in the energy while listing a range of potential policy tools. Such efforts have been further assisted by the increasing role of the European Court of Human Rights and case law under the Revised European Social Charter with its associated collective complaints mechanism. As stated in the CoE report,

> It is the Revised European Social Charter which gave a special emphasis to the housing problems of vulnerable social groups, which were reinforced by the CoE's Revised Strategy for Social Cohesion. (Council of Europe, 2008, p. 9)

The Third Energy Package and Vulnerable Consumers, 2009

The momentum behind the energy chapter in the Lisbon Treaty was also a driving force in the drafting of the TEP, which brings energy poverty and vulnerability into mainstream EU energy policy and establishes it as a European issue. The TEP consists of two directives—2009/72/EC on the internal market for electricity and 2009/73/EC on the internal market for natural gas—and three regulations concerning cross-border exchanges in electricity, transmission networks for natural gas and establishing the Agency for the Cooperation of Energy Regulators. In both directives (Paragraph 53 and Article 3 in Directive 2009/72/EC and paragraph 50 and Article 3 in Directive 2009/73/EC) an identical paragraph in the preamble and article in the main text identifies energy poverty as a 'growing problem' and requires action from Member States in two fields. Firstly, they must adopt a definition of the 'vulnerable consumer' and, secondly, they should ensure adequate protection for vulnerable consumers:

> Each Member State shall define the concept of vulnerable customers which may refer to energy poverty and, inter alia, to the prohibition of disconnection of gas to such customers in critical times. (Official Journal of the European Union, 2009, p. 211/103)

In line with the requirement to define vulnerable consumers in the TEP, Member States and national regulators have constructed country-specific characterizations of vulnerable consumers for use in national systems. The Council of European Energy Regulators (CEER) surveyed national regulators in 2012 and found that in 17 out of 26 countries the concept was included in energy laws or non-energy-sector laws, but cautioned that claiming not to have a formal definition does not mean that a given country does not have any protective measures in place (Council of European Energy Regulators, 2012). The review found that formal definitions range from targeting specific individuals to encompassing whole groups, but those countries without an explicit definition generally have as many protective measures as those with them. A similar review of the retail electricity market commissioned by Directorate General for Health and Consumers (SANCO) found that in Member States with a definition of 'energy poor' or 'non-affordable energy income threshold', the main criteria used are income thresholds, share of income required to meet adequate fuel requirements and consumer characteristics such as age and illness (Directorate General for Health and Consumers, 2010). The share of the population falling within the official definition of energy poor, though only established for three Member States, was found to vary between 6.6 and 18 per cent (ibid.).

While there is still no European consensus on what constitutes energy poverty, a 2010 Commission working paper suggested that those in energy poverty could be defined as 'households that spend more than a pre-defined threshold share of their overall consumption expenditure on energy products', where the threshold equals 'double of the national average ratio number' (European Commission, 2010a, pp. 10–16). A 2016 staff working document also carefully evaluates the EU's regulatory framework for electricity market design and consumer protection in the fields of electricity and gas, discussing existing Member State definitions of energy poverty and vulnerable consumers (European Commission, 2016).

A number of EU documents contain operational definitions which, while not specific to the energy sector, can form a basis upon which national definitions might be constructed. Directive 2005/29/EC on unfair practices in the internal market, for example, established a set of criteria that indicate how consumers might be vulnerable to unfair selling practices with regard to mental or physical infirmity, age and credulity. These general criteria covered a wide range of situations and were built upon in the European Parliament Resolution on strengthening the rights of vulnerable consumers, which examined extending the concept to 'include consumers in a situation of vulnerability, meaning consumers who are placed in a state of temporary powerlessness resulting from a gap between their individual state and characteristics on the one hand, and their external environment on the other hand' (European Parliament, 2012, p. 6). Directive 2009/73/EC itself states:

> Member States shall take appropriate measures to protect final customers, and shall, in particular, ensure that there are adequate safeguards to protect vulnerable customers. (Official Journal of the European Union, 2009, p. 211/103)

In its report on progress in the completion of the Internal Energy Market (IEM) in 2013, the European Council noted how many Member States have taken measures to protect vulnerable consumers and listed a number of examples: 'establishing a definition of vulnerable consumers for policy purposes, establishing suppliers of last resort, focused assistance, establishing social tariffs or social discounts, subsidized energy efficiency measures, setting minimum periods before energy deliveries can be suspended in cases of non-payment, prohibitions to cut off energy services during the winter, et cetera' (Council of the European Union, 2013, p. 9). A number of subsequent EU documents on the introduction of the TEP offered broad guidance on national measures that might be introduced, commonly emphasizing the need to use a policy mix that least distorts the IEM and is well targeted to reach only those truly in need of assistance.

The aforementioned SANCO Report found that most Member States used a combination of economic measures (to ensure prices are affordable and to assist consumers in arrears) and non-economic measures (regulation of the process for arrears and disconnection and support in finding the best tariff and increasing energy efficiency). The main types of measure are special prices offered to certain groups, helping to find the best tariff, energy-related payments (e.g. fuel allowance), grants to improve home energy efficiency and social security benefits (Directorate General for Health and Consumers, 2010). Similarly, the CEER found that Member States commonly employed specific protection measures for customers in remote areas, suppliers of last resort, default suppliers and social tariffs for vulnerable customers (Council of European Energy Regulators, 2012).

Despite having set a target to complete the IEM by 2014, however, the EU struggled to fully implement the provisions of the TEP. The Commission frequently pursued infringement proceedings against Member States not transposing or correctly transposing the TEP and its related legislation. Initially the Commission opened 38 infringement proceedings against 19 Member States, prompting an acceleration of national measures and the closure of many of the proceedings soon after. Additionally, the energy market was consistently identified as a problem in terms of lack of transparency and ease of consumer choice, even if offering considerable benefits in terms of competitiveness and growth. The policy focus on the IEM reflects the strength of the EU's mandate in internal market and consumer protection issues—while the Lisbon Treaty provides a comprehensive legal basis for EU energy policy action, the majority of initiatives and legislation have remained focused upon consumer rights and market liberalization, reflecting the early trajectory of the policy area.

The reliance upon IEM and consumer protection competence has produced an ongoing theme of energy as a public service, which runs throughout EU energy policy and is a key part of energy vulnerability policy. The TEP contains provisions on public service requirements—these capture the idea that energy is a vital public service and should strive for the highest standards of provision. Although these requirements are meant to be established at the national level, the EU advises that they should ensure fair pricing and competition, access to objective and transparent data, security of supply and mechanisms for the protection of vulnerable consumers.

The public service requirements build upon the idea of public service obligations (PSOs) included in early EU energy policy and highlighted in the Commission's 2007 Communication on an energy policy for Europe. In it, the Commission notes that PSOs do not go far enough and advocates instead for the launch of an ECC which should have as goal one to 'assist in establishing schemes to help the most vulnerable citizens deal with increases in energy prices' (Commission of the European Communities, 2007, p. 10). The ECC was eventually dropped—with its main provisions included in the TEP—but remained emblematic of the centrality of consumer protection in EU energy poverty and vulnerability policy.

Energy Efficiency and Energy Poverty, 2012

The second major source of energy poverty policy has been the EU's legislative framework on energy efficiency. This has evolved and developed alongside the general flux of energy policy but was most comprehensively captured in the 2012/27/EU EED. This document states in its preamble that national energy efficiency frameworks should ensure that vulnerable consumers have access to the benefits of high energy efficiency and notes the role of energy efficiency in reducing fuel poverty. Furthermore, Article 7 posits that within their energy efficiency targets Member States may 'include requirements with a social aim in the saving obligations they impose, including by requiring a share of energy efficiency measures to be implemented as a priority in households affected by energy poverty or in social housing' (European Commission, 2017c).

Historical energy efficiency policy in the EU has made little mention of energy vulnerability or fuel poverty, be it in relation to the energy performance of buildings, energy-using products or renewables; Directives 2002/91/EC, 2005/32/EC and 2006/32/EC establishing performance standards all lack provisions on energy vulnerability and poverty. In 2009, the directive on eco-design for energy-using products (Directive 2009/125/EC) was recast but still failed to mention energy vulnerability and poverty. This was only done in the third revision a year later, which notes the role of energy-efficient products in combatting energy poverty (see Directive 2010/31/EU). Finally, Commission guidance on the construction of national energy efficiency action plans has generally made little reference to energy poverty, even if recommending that action plans take into consideration the specificities of vulnerable consumers (European Commission, 2013b).

The International Energy Agency (IEA) identifies the difficulty of using energy efficiency regulation to tackle fuel poverty as the poor return on investment—often government spending—shown at evaluation. Most of the benefits are provided in the long term and are made not only to energy bills, but to tenants, property owners, energy providers, local communities and society as a whole, meaning that traditional cost–benefit analysis tends to misrepresent the return on initial capital. To address this and related issues, the IEA established a programme on innovative energy efficiency policies for mitigating fuel poverty, which seeks to take account of the wider benefits of energy efficiency as a tool for reducing fuel poverty (Heffner & Campbell, 2011).

Though there is considerable potential for the development of energy poverty policy through the EU's energy efficiency framework, this has been limited as policy has tended to prioritize climate change, commercial and environmental objectives over energy poverty goals. In the long term, many of the aims of climate change and energy poverty policy are entirely compatible and mutually reinforcing. But in the short term, policy tools for addressing energy poverty have tended to involve financial aid as part of national social policy and energy pricing policy, serving to relieve the financial burden rather than reducing energy consumption via more efficient buildings and products. Meanwhile, short-term climate change policies tend to increase costs for the end user by imposing higher regulatory standards and prices. A Commission Green Paper noted the 'impact on energy prices, adversely impacting affordability of energy for vulnerable households and the competitiveness of energy intensive sectors even though they may reduce industry's exposure to energy costs and improve resilience to energy price peaks', while the 2020 climate change goals make no reference to energy poverty (European Commission, 2013a, p. 10).

Social Policy, Economic Crisis and the Vulnerability 'Policy Mix'

The third main source of EU energy vulnerability policy has been social policy; though in itself this is a misnomer, since social policy is a largely national competence. Social policy measures to combat energy poverty typically take the form of financial assistance—social tariffs, lower tariffs of subsidies made available to vulnerable consumers, such as winter and cold weather payments for the elderly or disabled (European Commission, 2013c). A small but consistent thread of reference to the social dimension of energy policy exists throughout its evolution and, in the wake of the economic crisis and concerns about rising costs and vulnerable populations, this has taken on a new significance.

At the launch of the Europe 2020 Strategy in 2010, the European Commission noted that 'to lift people out of poverty will require access to energy since achieving the goal of eradicating extreme poverty by 2015 cannot be met unless substantial progress is made on improving access' (European Commission, 2010b, p. 17). This was the first statement linking energy to poverty eradication within the EU context, as opposed to external relations and development. It was soon followed by a number of similar policy statements. During the discussion of the 2050 Energy Roadmap, the Commission stated that 'as energy poverty is one of the sources of poverty in Europe, the social aspects of energy pricing should be reflected in the energy policy of Member States' (European Commission, 2011, p. 17), while the Parliament welcomed the inclusion of a social dimension and insisted that the issue of energy poverty be given special attention.

An important social policy statement on energy poverty was made by the European Economic and Social Committee (EESC), which highlighted the importance of links with other sectors, such as health, consumer rights and housing, urging Member States to do more to combat energy poverty and calling on the Commission to establish a European Energy Poverty Monitoring Centre to provide better research and facilitate the mainstreaming of energy poverty into other policy areas (European Economic and Social Committee, 2011). The EESC asserted:

> The European Union legislates on energy policy, has powers in this field and consequently has an impact, whether direct or indirect, on energy poverty in the Member States. The EU must, therefore, act and deliver policies within its sphere of competence. (European Economic and Social Committee, 2011, p. 44/56)

Drawing on the EESC's calls for an intersectoral approach to energy poverty, the European Parliament included a dedicated section on combatting energy poverty in its 2012 Resolution on social housing (European Parliament, 2013b). This called for energy-efficiency standards to be incorporated into definitions of 'decent housing' and stated that access to energy should be considered a requirement in order for people to lead a 'dignified life'. The Parliament called on Member States to enact a series of measures to combat energy poverty including financial schemes and regional funds to assist vulnerable consumers, programmes of incentives and instructive measures to help residents reduce energy consumption and long-term financial leverage instruments to make buildings more energy efficient. In drawing a link to public health, the Resolution noted that measures to reduce energy poverty can help to tackle illnesses including 'respiratory and cardiovascular conditions, allergies, asthma, food and carbon monoxide poisoning, and impacts on mental health'. Emphasizing the need to utilize energy, consumer and social policy to tackle energy poverty, the Council called on the Commission to provide a follow-up review on these issues by 2016 (Council of the European Union, 2014b).

The post-2008 financial crisis and economic recession prompted Member States and the EU institutions to look closer at national expenditure. A 2014 report on energy prices and costs in Europe drew a link between the economic crisis, rising prices and energy poverty, noting that 'the on-going financial and economic crisis makes addressing energy poverty and/or vulnerability more important today, given that energy cost rises are hitting poor households harder' (European Commission, 2014b, p. 14). European Council conclusions have confirmed a similar focus, discussing high energy costs and affordable energy prices as priorities in the completion of the IEM, also calling for 'sustained efforts to moderate the energy costs borne by energy end-users' (European Commission, 2014c, p. 1). Also, the Council of the EU, in its discussions on the rising cost of energy in Europe, has urged Member States to use cost-effectiveness and price contestability to protect both vulnerable consumers and competitiveness.

Revolutionizing EU Energy Poverty Policy: The Clean Energy Package

In recent years, the European Commission has taken a further lead in advancing the EU energy poverty agenda. This has been primarily achieved via the co-ordination of a project aimed at investigating the policies and measures currently in place to protect 'vulnerable consumers' in the energy sector across the EU (Pye et al., 2015). In addition to discussing the multiple meanings and understandings of vulnerability, this document emphasizes that many measures are being implemented across Member States, focused both on vulnerable consumers and on energy poverty. However, these are distinct issues, and are targeted by different types of measures. Measures focused on vulnerable consumers offer protection within regulated markets, and facilitate access and participation. They are often short-term in nature, providing relief or ensuring on-going supply in the face of indebtedness. Energy poverty measures on the other hand are explicitly focused on lower income households, and seek to address longer term structural problems of building energy efficiency. (Pye et al., 2015, p. vii)

A further Commission-sponsored report (Rademaekers et al., 2016) has dealt with the identification of ways to measure the problem, proposing four key indicators tested and computed for the Netherlands, Slovakia, Spain and Italy using currently available data. The Commission has also supported the publication of an extensive analysis of the components of energy prices and costs in Europe, with an emphasis on household energy budgets among different income groups (Grave et al., 2016). This document points to the inflationary character of taxes, levies and network charges on final energy prices in Europe, while underlining that the main impact of increasing retail prices is on low-income households:

In some countries, increasing average energy costs have been compensated by increasing average income, thus, mitigating price effects. However, low income households have high shares of residential energy expenditures and are affected most by changes in energy retail prices. A number of countries introduced policies to support households with low income to keep their standards of living, either through issuing allowances to cover costs for heating, or by reduced tariffs per unit of energy. (Grave et al., 2016, p. 6)

As a whole, this body of work has helped establish the state of the art in terms of policy knowledge and the measurement of energy poverty, while further affirming the importance of energy poverty as a decision-making concern across the European context. The three reports have identified the diversity of measures, drivers and circumstances that can be attributed to energy poverty, bringing out the difficulties associated with capturing the problem via a single indicator. They have pointed to a lack of consistency in the understanding and treatment of energy poverty at the level of Member States—with some countries treating the issue as a social policy challenge, and others developing a more comprehensive approach:

It would also be possible to further improve the data available from existing surveys. As mentioned in the report, while there have been efforts for harmonization, differences remain in frequency, timing, content and structure of the Household Budget Surveys. A number of recommendations are available to improve datasets at the EU level to further enhance the measurement of energy poverty. One specific simple recommendation is to include a variable in the SILC survey that refers to the total energy spending. If this were the case, it would be possible to calculate all energy poverty metrics from this survey. (Rademaekers et al., 2016, p. 98)

The 'Clean Energy for all Europeans' package has allowed for the materialization of these analytical documents into policy stipulations and legal regulation. The Package is made up of eight legislative proposals targeting a variety of sectors, inlcuding energy efficiency, renewable energy, electricity market redesign, governance rules for the Energy Union, energy security and eco-design. The proposals are currently being discussed within the European Council and European Parliament with a view to being adopted in 2018. Their departure point is a strong declarative commitment to energy efficiency, 'fair treatment' of consumers and global leadership in energy transitions.

A number of the Clean Energy Package proposals explicitly mention energy poverty. For example, the draft of the newly 'recast' Electricity Directive (European Commission, 2017b) maintains the Article 28 Directive 2009/72/EC provision on common rules for the internal market in electricity, which requires Member States to 'define the concept of vulnerable customers which may refer to energy poverty and, inter alia, to the prohibition of disconnection of electricity to such customers in critical times'. In Article 5 of the recast Directive, the European Commission also obliges Member States to ensure protection of energy poor or vulnerable customers 'in a targeted manner by other means than public interventions in the price-setting for the supply of electricity'. The same draft Directive also provides a definition of energy poverty:

> Energy poor households are unable to afford these energy services due to a combination of low income, high energy expenditure and poor energy efficiency of their homes. (recital 40)

The new Electricity Directive proposes that Member States should define a set of criteria to measure energy poverty, while being obliged to monitor the number of households in energy poverty so as to provide targeted support. In Article 29, it is suggested that Member States 'shall report on the evolution of energy poverty and measures taken to prevent it to the Commission every two years as part of their Integrated National Energy and Climate Progress Reports'.

Also of importance are amendments to the 2012 EED. One of these builds on a key provision stating that Member States 'may include requirements with a social aim in the saving obligations they impose … by requiring a share of energy efficiency measures to be implemented as a priority in households affected by energy poverty or in social housing' (European Commission, 2012, p. 315/16). In the amendments, the provision is modified by replacing 'may' with 'shall'. The new EED also introduces a new requirement whereby 'in designing alternative policy measures to achieve energy savings, Member States shall take into account the effect on households affected by energy poverty' (European Commission, 2017c, p. 20). Also, the draft Energy Performance of Buildings Directive states:

> This proposal could contribute to taking out from energy poverty between 515 000 and 3.2 million households in the EU (from a total of 23.3 million households living in energy poverty—Eurostat). (European Commission, 2017a, p. 3)

The preamble to the same draft Directive states that the social impacts of energy building improvements will also be felt in the energy poverty domain, while stipulating that the European Commission will require Member States to contribute to the alleviation of energy poverty through their long-term renovation strategy (Article 2).

Last but not the least, Article 21 in the Governance of the Energy Union Regulation affirms that the Member State–level Integrated National Energy and Climate Progress Reports will include information on the implementation of 'national objectives with regards to energy poverty, including the number of households in energy poverty' (European Commission, 2017d, p. 38). The provision of alternative measures for energy savings—such as those relevant to social housing—integrated within the EED are also mentioned.

While it remains unclear, at the time of writing this book, to what extent these extensive provisions will be translated and implemented into binding legal documents, there is little doubt that energy poverty now plays a much more pivotal role as a pan-EU concern. Aside from the pioneering role of the European Commission in moving this agenda forward, recent activities in other institutions, notably the European Parliament, have also played a role. This involves the increased frequency and extent of parliamentary debates as well as engagement of specific parliamentary committees: on Industry, Research and Energy; Employment and Social Affairs; and Women's Rights and Gender Equality. Widely cited across policy and scientific circles has been a dedicated European energy poverty handbook (Csiba, 2016) and video (https://youtu.be/0tZ9-9hmSOw) published by the Greens/European Free Alliance group; while the Socialists and Democrats adopted a manifesto to 'fight energy poverty' in the context of the Energy Union (S&D Manifesto, 2017).

Beyond formal political institutions, there is also an ever-expanding polity of European-level think tanks and industrial lobby groups who have published position papers or analytical work on energy poverty. Notable examples include a policy brief issued by the Union of the Electricity Industry highlighting, inter alia, that 'any new policy initiative at EU level should be subject to a distributional impact assessment to make sure that energy customers—especially the most vulnerable ones—will not bear disproportionate risks and unintended consequences' (Eurelectric, 2017, p. 6). The eminent Jacques Delors Institute has also commented on energy poverty (Pellerin-Carlin, Vinois, Rubio, & Fernandes, 2017) arguing that Europe needs 'a social pact for the energy transition' as well as a 'European action plan to eradicate energy poverty'. The Social Platform has also expressed an interest in the issue (Bouzarovski, 2014), as has the European Policy Centre (Dhéret & Giuli, 2017).

Energy Poverty Governance: A Hybrid Model

The evolution of EU energy poverty policy has been directed, for a significant part, by the evolution of the EU's governance system and legal mandates. As a result, one can observe the development of a hybrid model of governance in energy poverty policy, supplementing hard law in strong-mandate areas with soft law in those areas where Member States retain autonomy.

The central threads of the EU energy poverty policy have been the IEM and the protection of vulnerable consumers. This is not coincidental, but rather reflects the competence assigned to the EU as a supranational body. Though energy poverty is understood as a social problem and the role of social policy in addressing its causes is widely acknowledged, the Commission continues to legislate and project within the bounds of its constitutional asymmetry, elevating the consumer and the proper functioning of the IEM as the focus of policy. Though the changing economic environment has elevated the need to protect vulnerable members of society, social, health and welfare policy remain largely national responsibilities, leaving the EU to tackle issues such as energy poverty from one side—legislating on the internal market and consumer protection while leaving social policy remedies to the Member States.

To get around this imbalance, the EU supplements this well-established internal market and consumer legislation with a vast body of soft law. In 2007, the Commission established the Citizens' Energy Forum (CEF), a platform designed to implement and enforce consumer rights in the energy market, bringing together national consumer organizations, industry, national regulators and government authorities. Initially working on issues such as smart metres, user-friendly billing and switching suppliers, the CEF established a new working group in 2011, commencing activities in early 2012, to examine the policy framework for the protection of vulnerable consumers. This Vulnerable Consumers Working Group (VCWG), established by DG Energy in close collaboration with SANCO, has had a number of aims and objectives, including to review factors that impact consumers' energy poverty, to assess the drivers of vulnerability, to develop key characteristics of vulnerable consumers and what differentiates them, to consider energy policy and non-energy instruments that can address vulnerability.

The VCWG's activities have included collecting and assessing illustrative existing practices, reviewing data and generating recommendations for action. For the most part, Member States have supported this emphasis upon voluntary collaboration. The Council has endorsed the use of benchmarking and exchange of best practice as appropriate mechanisms for coordination, commonly calling on the Commission to presents 'reviews' or 'reports' rather than legislative solutions. Crucially, Member States agreed in 2015 to the systematic monitoring of key indicators for an 'affordable, safe, competitive, secure and sustainable energy system' (European Council, 2015). This presents a potential foundation for ongoing monitoring of the core factors affecting energy poverty from within the energy sector, to complement current data sources related to income, social housing provision and 'at risk of poverty or social exclusion' status.

The 2014 European Commission's Communication on energy prices and costs advises Member States that 'for households, fiscal transfers can be considered to provide protection, bearing in mind that it is generally more efficient to protect such vulnerable consumers through social policy measures (such as fiscal transfers) rather than through energy pricing' (European Commission, 2014b, p. 243). Similar conclusions about the optimal policy mix and the risks of using energy pricing as a lever are made by the Council and in the TEP. Yet the staff working document that accompanies the Commission's 2014 Communication stated that 'social tariffs may distort the market, do not encourage energy-efficient behaviour, and have a proportionally higher financial impact on those who fall just outside the vulnerable classification' (European Commission, 2014a, p. 243).

This shift in the recommended policy mix is also reflected in policy statements which encourage the use of 'coordinated and balanced social, energy and consumer policy, as determined by each Member State to combat energy poverty' (Council of the European Union, 2014a, p. 5). Earlier references mentioned primarily energy and social policy, while noting clearly that the latter remained a matter of national competence. The changing perception of 'best policy mix' for combatting energy poverty can thus be said to reflect both the growing EU role in this area and the economic circumstances in which it operates.

Trends and Developments in EU Energy Poverty Policy

Energy poverty policy in the EU has evolved along broadly the same path as energy policy—virtually non-existent until the mid-2000s it has become a consistent component of EU policy despite the shared competences that dictate its governance. Its primary source remains the IEM and the EU's considerable body of policy on consumer protection in the common market. Though promising indications were seen in the EED, these have not translated into a solid second source of policy, largely because of the inherent conflict with climate change and environmental objectives, particularly in the short term. More recently, the relevance of social policy has been acknowledged but the subsidiarity principle and the absence of a social policy mandate prevent the EU from developing comprehensive policy on this basis. That said, an early focus upon vulnerable consumers has shifted somewhat, since the onset of the economic recession, to the social dimension of energy policy and the need to combat energy poverty with a coordinated, cross-sector approach.

Evaluating the success of specific policies on energy poverty is difficult; data is lacking and many policy instruments are non-binding and flexible, meaning that Member States may choose to implement them in a number of different ways. The TEP, which required that Member States adopt a definition of vulnerable consumers and take measures to ensure that they are protected, suffered initial difficulties in implementation, resulting in the opening of a number of infringement proceedings. But its provisions relating to energy poverty are generally considered to have been well transposed—for the most part, Member States already had implicit notions of vulnerable consumers in national law and almost all have some form of measure to protect those who qualify. As such, the TEP might be deemed a 'success' in that it has raised the profile of energy poverty as an EU issue and performed a key role in laying the foundation for common discussion. Energy efficiency legislation is hard to evaluate from an energy poverty perspective because of the non-binding language in which it is couched; although this is now being gradually changed by the provisions of the Clean Energy Package. EU-level policy statements on the use of social policy instruments to combat energy poverty are vague, non-binding and, as yet, are not consolidated into a coherent policy objective.

More broadly, the 'success' of EU energy poverty initiatives might be judged by asking to what extent they have become a genuine EU policy. From this perspective, though fragmented in its approach, the EU has moved to some extent in establishing energy poverty as a European issue and laying the foundations for a coherent policy in this area. The TEP made important steps in mainstreaming energy poverty into energy policy, though some lags have been experienced in energy efficiency, for example. Since this initial introduction, almost all subsequent policy on the IEM has made reference to energy poverty as a component and policy objective, though with varying specificity and force. Furthermore, repeated emphasis of the link to other sectors, such as health, consumers and housing, has resulted in a widely recognized need for a coordinated approach in tackling energy poverty. European-level platforms have been welcomed by Member States and stakeholders and have produced valuable reviews of existing practice and recommendations for further action, as well as highlighting the necessity and benefit of a common EU approach. As such, and in spite of its imbalanced competence in the relevant areas, the EU has succeeded in taking the first steps to making energy poverty a European policy issue.

Actors and Stakeholders in EU Energy Poverty Policy

The driving actor in pushing an EU energy poverty policy forward has been the European Commission. Following numerous failed attempts to create a common energy policy framework prior to 2007, the Commission has ensured that energy poverty forms an important strand of the now wide-ranging EU action in energy. As in all areas of EU policy, the Commission acts as the primary agenda-setter in energy policy, initiating and drafting legislation but restrained in its financial and administrative resources, as well as its legal mandate (Birchfield, 2011). The individual DGs are also important for expanding and developing energy poverty policy and have been relatively effective in framing and defining it as a problem which should be dealt with by the DGs for environment, health, consumers and, to some extent, foreign relations, in addition to DG Energy. Implementation of the various instruments and legislation has also been supported by DG Competition, which has stepped in when Member States have not fully transposed IEM rules or implemented other energy regulations. Support with enforcement has also been received from the Court of Justice in its role as adjudicator on infringement proceedings brought against national governments.

While the Commission has carefully balanced its energy policy goals with its constitutional asymmetry and need to achieve sufficient member state support, other EU institutions have been outspoken about the necessity of a coherent EU energy poverty policy. Both the European Parliament and the EESC have made bold statements about the social dimension of energy poverty, the pan-European nature of the problem, the lack of sufficient action on the part of Member States and the need for the EU to exercise its indirect influence in spite of its competence weakness. This impetus for greater EU involvement is supported by the research and resources gathered by the various platforms and stakeholder groups at the EU level. Bodies such as the CEF and the VCWG have been instrumental in making available the kind of background research on existing practice and areas for improvement called for in early energy poverty policy statements, thus paving the way for further action; and energy poverty has received frequent attention at high-level fora such as EU Sustainable Energy Week (Fig. 3.1).

This driving support and impetus for action is weathered by the somewhat more passive role of Member States. Reluctant to cede any more responsibility in the energy sector than is necessary for the functioning of the Single Market, the various configurations of the Council have been careful to moderate the momentum of Commission activity, preferring in most cases to use soft law mechanisms and non-binding measures to pursue common objectives, rather than legislative solutions.

Consequently, energy poverty policy has an 'emergent and precarious nature', lacking an institutional centre and limited by its dependence upon the consumer protection agenda, which prevents it from tackling the structural conditions which cause energy and fuel poverty (Bouzarovski et al., 2012).

Fig. 3.1 An energy poverty session at European Energy Week, 2017 (photo by Saska Petrova)

Historically, an EU-wide definition of energy poverty or vulnerable consumers has been deliberately avoided by the institutions because of the difficulty in designing a concept which fits with all existing national understandings. While the decision to refrain from adopting a common definition aimed to encourage Member State action by maximizing flexibility, this is identified in much of the literature as the fundamental cause of the insufficient measures taken to date and the primary barrier to more coherent EU action (Thomson et al., 2016). Different countries also require different policy mixes and measures to address energy poverty. An expert workshop held in Brussels in 2010 raised concerns that policies led by the EU showed insufficient consideration for the lack of institutional capacity to deal with energy poverty at multiple levels of governance (Bouzarovski et al., 2012); similar issues have been discussed at subsequent events (Fig. 3.2). In light of the vague and impractical nature of policy recommendations made at the EU level, the group concluded that very little direct action has therefore been taken at the different levels, and is unlikely to be taken unless a common definition and an established evidence base can be provided.

Fig. 3.2 Participants at an international conference on energy poverty and vulnerability in Manchester, in 2013 (photo by Stefan Bouzarovski)

Instead of a Conclusion: The Future of the EU Energy Poverty Policy

The evolution of EU energy poverty to date, as tempered by the imbalance of relevant competences and disparate policy tools available, indicates that future policy development is likely to proceed along the established path of supplementing peripheral hard law with soft law instruments. But an often overlooked avenue for the development of energy poverty policy is via the EU's health agenda. Health actors have consistently taken an interest in energy poverty, and the link between lack of access to energy and poor mental and physical health in people of all ages is well established (Heffner & Campbell, 2011). The EU's public health mandate is relatively strong; and recent research commissioned in light of the economic recession and social crisis has highlighted the relevance of energy poverty in Europe's recovery and return to growth. The health community has also been instrumental in forging the link between energy poverty and housing quality, supported by a coalition of stakeholders concerned with the quality of the housing stock and the implementation of the EED.

Technology and climate change are two other strategic directions likely to affect the direction of the EU energy poverty policy in the coming years. Smart metering—a growing priority of EU energy policy—holds the potential to combat energy poverty in some instances, but its pros and cons for energy-poor users require careful consideration (Darby, 2012). Low-carbon urban and regional development policies more generally also hold significant energy poverty reduction opportunities, especially if justice contingencies are taken into account. There are also important intersections between climate change policy and energy poverty policy—not only in terms of mitigation efforts but also in relation to the impacts of global warming on the need for additional energy services in the home, such as space cooling. These complex policy loci are at odds with the relatively unidirectional nature of EU energy poverty initiatives to date: in Bouzarovski and Petrova (2015b), we noted that agenda-shaping in the EU poverty domain has been mainly driven from above, and has been highly contingent on attempts to 'define' and 'identify' the problem. It remains to be seen whether the significant—and relatively fast—development of a distinctive EU agenda on energy poverty will be accompanied by a range of groundbreaking steps to capture the more systemic implications of the problem.

REFERENCES

Andoura, S., Hancher, L., & van der Woude, M. (2010). *Towards a European Energy Community: A policy proposal.* Brussels: Notre Europe.
Bakker, K. (2007). The 'commons' versus the 'commodity': Alter-globalization, anti-privatization and the human right to water in the global South. *Antipode, 39,* 430–455.
Birchfield, V. (2011). The role of EU institutions in energy policy formation. In V. Birchfield & J. S. Duffield (Eds.), *Toward a common European Union energy policy* (pp. 235–262). New York: Palgrave Macmillan.
Bouzarovski, S. (2014). *Social justice and climate change: Addressing energy poverty at the European scale.* Brussels: Spring Alliance.
Bouzarovski, S., & Petrova, S. (2015a). A global perspective on domestic energy deprivation: Overcoming the energy poverty–fuel poverty binary. *Energy Research & Social Science, 10,* 31–40.
Bouzarovski, S., & Petrova, S. (2015b). The EU energy poverty and vulnerability agenda: An emergent domain of transnational action. In J. Tosun, S. Biesenbender, & K. Schulze (Eds.), *Energy policy making in the EU* (pp. 129–144). London: Springer.
Bouzarovski, S., Petrova, S., & Sarlamanov, R. (2012). Energy poverty policies in the EU: A critical perspective. *Energy Policy, 49,* 76–82.
Commission of the European Communities. (2005). *Staff working document 'Annex to the Communication from the Commission "Annual report 2005 on the European Community's development policy and the implementation of external assistance in 2004"'.* Brussels: Commission of the European Communities.
Commission of the European Communities. (2007). *Communication 'An energy policy for Europe'.* Brussels: Commission of the European Communities. Retrieved from http://eur-lex.europa.eu/legal-content/EN/TXT/PDF/?uri=CELEX:52007DC0001&from=EN
Council of Europe. (2008). *Housing policy and vulnerable social groups.* Strasbourg: Council of Europe.
Council of European Energy Regulators. (2012). *Status review of customer and retail market provisions from the 3rd package as of 1 January 2012.* Brussels: Council of European Energy Regulators.
Council of the European Union. (2007). *Presidency conclusions.* Brussels: Council of the European Union.
Council of the European Union. (2013). *Progress on the completion of the Internal Energy Market.* Brussels: General Secretariat of the Council.
Council of the European Union. (2014a). *Completion of the Internal Energy Market.* Brussels: Council of the European Union.
Council of the European Union. (2014b). *Council conclusions on 'Energy prices and costs, protection of vulnerable consumers and competitiveness'.* Brussels:

Council of the European Union. Retrieved from http://www.consilium.europa.eu/uedocs/cms_data/docs/pressdata/en/trans/143198.pdf

Csiba, K. (Ed.). (2016). *Energy poverty handbook*. Brussels: European Parliament.

Darby, S. J. (2012). Metering: EU policy and implications for fuel poor households. *Energy Policy, 49,* 98–106.

Dhéret, C., & Giuli, M. (2017). *The long journey to end energy poverty in Europe.* Brussels: European Policy Centre.

Directorate General for Health and Consumers. (2010). *The functioning of retail electricity markets for consumers in the European Union.* Brussels: ECME Consortium. Retrieved from http://ec.europa.eu/consumers/archive/consumer_research/market_studies/docs/retail_electricity_full_study_en.pdf

Dubois, U., & Meier, H. (2016). Energy affordability and energy inequality in Europe: Implications for policymaking. *Energy Research & Social Science, 18,* 21–35.

Duffield, J., & Birchfield, V. (2011). Introduction: The recent upheaval in EU energy policy. In J. Duffield & V. Birchfield (Eds.), *Toward a common European Union energy policy* (pp. 1–12). New York: Palgrave Macmillan.

Eurelectric. (2017). *Energy poverty. A Eurelectric position paper.* Brussels: Union of the Electricity Industry—Eurelectric.

European Anti-Poverty Network. (2017). *Right to energy for all Europeans.* http://www.eapn.eu/wp-content/uploads/2017/06/EAPN-2017-letter-to-MEPs-Right-to-Energy-Coalition-1225.pdf. Retrieved September 1, 2017.

European Commission. (2010a). *Commission staff working paper: An energy policy for consumers.* Brussels: EC.

European Commission. (2010b). *Communication 'Energy 2020 a strategy for competitive, sustainable and secure energy'.* Brussels: European Commission.

European Commission. (2011). Communication from the European Commission to the European Parliament, the Council, the European Economic and Social Committee and the Committee of the Regions. *Energy roadmap 2050.* Brussels: European Commission. Retrieved from http://eur-lex.europa.eu/LexUriServ/LexUriServ.do?uri=COM:2011:0885:FIN:EN:PDF

European Commission. (2012). *Directive 2012/27/EU of the European Parliament and of the Council of 25 October 2012 on energy efficiency, amending Directives 2009/125/EC and 2010/30/EU and repealing Directives 2004/8/EC and 2006/32/EC Text with EEA relevance.* http://eur-lex.europa.eu/legal-content/EN/TXT/?uri=celex:32012L0027. Retrieved 1st September 2017.

European Commission. (2013a). *Green Paper 'A 2030 framework for climate and energy policies'.* Brussels: European Commission. Retrieved from http://eur-lex.europa.eu/legal-content/EN/TXT/PDF/?uri=CELEX:52013DC0169&from=EN

European Commission. (2013b). *Staff working document 'Guidance for national energy efficiency action plans'.* Brussels: European Commission.

European Commission. (2013c). *Vulnerable consumer working group guidance document on vulnerable consumers, November 2013*. Brussels: European Commission, Vulnerable Consumer Working Group.
European Commission. (2014a). *Commission staff working document energy prices and costs report*. Brussels: European Commission.
European Commission. (2014b). Communication from the Commission to the European Parliament, the Council, the European Economic and Social Committee and the Committee of the Regions. *Energy prices and costs in Europe*. Brussels: European Commission.
European Commission. (2014c). Communication from the Commission to the European Parliament, the Council, the European Economic and Social Committee and the Committee of the Regions. *Green Action Plans for SMEs*. Brussels: European Commission.
European Commission. (2015). Communication from the Commission to the European Parliament, the Council, the European Economic and Social Committee and the European Investment Bank. *A Framework Strategy for a Resilient Energy Union with a Forward-Looking Climate Change Policy*. Brussels: European Commission. Retrieved from http://eur-lex.europa.eu/legal-content/EN/TXT/?uri=COM:2015:80:FIN
European Commission. (2016). *Commission staff working document 'Evaluation report covering the evaluation of the EU's regulatory framework for electricity market design and consumer protection in the fields of electricity and gas'*. Brussels: European Commission.
European Commission. (2017a). *Proposal for a directive of the European Parliament and of the Council amending Directive 2010/31/EU on the energy performance of buildings*. http://eur-lex.europa.eu/legal-content/EN/TXT/?uri=CELEX: 52016PC0765. Retrieved September 1, 2017.
European Commission. (2017b). *Proposal for a directive of the European Parliament and of the Council on common rules for the internal market in electricity (recast)*. http://eur-lex.europa.eu/legal-content/EN/TXT/?uri=CELEX%3A52016P C0864R%2801%29. Retrieved September 1, 2017.
European Commission. (2017c). *Proposal for a directive of the European Parliament and of the Council on energy efficiency and repealing Directives 2004/8/EC and 2006/32/EC*. http://eur-lex.europa.eu/legal-content/en/ALL/?uri=CELEX: 52011PC0370. Retrieved September 1, 2017.
European Commission. (2017d). *Proposal for a Regulation of the European Parliament on the Governance of the Energy Union, amending Directive 94/22/EC, Directive 98/70/EC, Directive 2009/31/EC, Regulation (EC) No 663/2009, Regulation (EC) No 715/2009, Directive 2009/73/EC, Council Directive 2009/119/EC, Directive 2010/31/EU, Directive 2012/27/EU, Directive 2013/30/EU and Council Directive (EU) 2015/652 and repealing Regulation (EU) No 525/2013*. http://eur-lex.europa.eu/legal-content/EN/TXT/?uri=COM:2016:759:REV1. Retrieved September 1, 2017.

European Council. (2015). *Council conclusions on the governance system of the Energy Union*. http://www.consilium.europa.eu/en/press/press-releases/2015/11/26-conclusions-energy-union-governance/. Retrieved September 1, 2017.

European Economic and Social Committee. (2011). *Opinion of the European economic and social committee on 'Energy poverty in the context of liberalisation and the economic crisis' (exploratory opinion)*. Brussels: European Economic and Social Committee.

European Group on Ethics in Science and New Technologies to the European Commission. (2013). *Opinion no 27 'An ethical framework for assessing research, production and use of energy'*. Brussels: European Commission.

European Parliament. (2012, May 22). *Resolution 'strengthening the rights of vulnerable consumers'*. Retrieved from http://www.europarl.europa.eu/sides/getDoc.do?pubRef=-//EP//NONSGML+TA+P7-TA-2012-0209+0+DOC+PDF+V0//EN

European Parliament. (2013a). *European Parliament resolution of 14 March 2013 on the Energy roadmap 2050, a future with energy*. http://www.europarl.europa.eu/sides/getDoc.do?pubRef=-//EP//TEXT+TA+P7-TA-2013-0088+0+DOC+XML+V0//EN. Retrieved September 1, 2017.

European Parliament. (2013b). *Report on social housing in the European Union*. http://www.europarl.europa.eu/sides/getDoc.do?pubRef=-//EP//NONSGML+REPORT+A7-2013-0155+0+DOC+PDF+V0//EN. Retrieved September 1, 2017.

Grave, K., Breitschopf, B., Ordonez, J., Wachsmuth, J., Boeve, S., Smith, M., ... Schleich, J. (2016). *Prices and costs of EU energy*. Utrecht: Ecofys Netherlands.

Heffner, G., & Campbell, N. (2011). *Evaluating the co-benefits of low-income energy-efficiency programmes*. Paris: International Energy Agency.

Hiteva, R. P. (2013). Fuel poverty and vulnerability in the EU low-carbon transition: The case of renewable electricity. *Local Environment, 18*, 487–505.

Intelligent Energy Europe. (2017). *European fuel Poverty and Energy Efficiency (EPEE)*. https://ec.europa.eu/energy/intelligent/projects/en/projects/epee. Retrieved September 1, 2017.

McCormick, J. (2014). *Understanding the European Union: A concise introduction*. London: Palgrave Macmillan.

Official Journal of the European Union. (2009). *Directive 2009/73/EC of the European Parliament and of the Council of 13 July 2009 concerning common rules for the internal market in natural gas and repealing Directive 2003/55/EC (Text with EEA relevance)*. http://eur-lex.europa.eu/legal-content/EN/ALL/?uri=CELEX:32009L0073. Retrieved September 1, 2017.

Official Journal of the European Union. (2012). *Consolidated version of the Treaty on the Functioning of the European Union*. http://eur-lex.europa.eu/legal-content/EN/TXT/?uri=CELEX:12012E/TXT. Retrieved September 1, 2017.

Pellerin-Carlin, J., Vinois, J.-A., Rubio, E., & Fernandes, S. (2017). *Making the energy transition a European success: Tackling the democratic, innovation, financing and social challenges of the energy union*. Paris: Jacques Delors Institute.

Pye, S., Baffert, C., Brajković, J., Grgurev, I., Miglio, D. R., & Deane, P. (2015). *Energy poverty and vulnerable consumers in the energy sector across the EU: Analysis of policies and measures*. London: Insight_E.

Rademaekers, K., Yearwood, J., Ferreira, A., Pye, S., Hamilton, I., Agnolucci, P., ... Anisimova, N. (2016). *Selecting indicators to measure energy poverty*. Brussels: European Commission, DG Energy.

S&D Manifesto. (2017). *Fighting energy poverty—S&D Manifesto*. http://www.socialistsanddemocrats.eu/publications/fighting-energy-poverty-sd-manifesto. Retrieved September 1, 2017.

Thomson, H., Snell, C., & Liddell, C. (2016). Fuel poverty in the European Union: A concept in need of definition? *People Place and Policy, 10,* 5–24.

Walker, G. (2015). The right to energy: Meaning, specification and the politics of definition. *L'Europe en Formation, 378*(4), 26–38.

Open Access This chapter is distributed under the terms of the Creative Commons Attribution 4.0 International License (http://creativecommons.org/licenses/by/4.0/), which permits use, duplication, adaptation, distribution and reproduction in any medium or format, as long as you give appropriate credit to the original author(s) and the source, a link is provided to the Creative Commons license and any changes made are indicated.

The images or other third party material in this chapter are included in the work's Creative Commons license, unless indicated otherwise in the credit line; if such material is not included in the work's Creative Commons license and the respective action is not permitted by statutory regulation, users will need to obtain permission from the license holder to duplicate, adapt or reproduce the material.

CHAPTER 4

The European Energy Divide

Abstract This chapter reviews the spatial and social differences that underpin existing and past patterns of energy poverty in Europe. This is achieved via exploration of scientific research focused on the topic, either as a central object of enquiry or as part of wider investigations in which the issue is brought up as a relevant factor. Special attention is paid to the large-scale geographic variation of energy poverty in Europe, as well as the manner in which this diversity is subsequently reflected at the level of nations, regions and particular demographic groups. I also discuss the driving forces of energy poverty within particular spatial contexts; and in an effort to move beyond the traditional geographic focus of energy poverty research—the UK and Ireland—the chapter first systematically overviews the development of a debate focusing on different parts of the European continent and its immediate neighbourhood. I then review evidence about patterns of energy poverty at a variety of spatial scales.

Keywords Energy poverty • Energy vulnerability • Material deprivation • Uneven development • Europe

© The Author(s) 2018
S. Bouzarovski, *Energy Poverty*,
https://doi.org/10.1007/978-3-319-69299-9_4

INTRODUCTION

As was noted earlier in the book, the majority of evidence about the underlying causes of energy poverty has been generated by studies undertaken in the UK and the Republic of Ireland (RoI). Academic research on 'fuel poverty' produced in these two states has uncovered that the condition is brought about, in the main, by the interaction of low household incomes with thermally inefficient homes (Boardman, 2010). It has been underlined that the residents of inefficient dwellings are forced to purchase less affordable energy services than the rest of the population, because such homes are more expensive to heat. In relative terms, energy services are also less affordable to income-poor households, since such families will have lower amounts of disposable funds for such purposes. But the fact that fuel poverty is co-produced by energy efficiency and low incomes means that not all income-poor households will also be fuel poor. Additionally, the extensive nature of fuel poverty in the UK and RoI—itself a product of the two countries' specific inequality patterns and housing stock structure—has allowed for an additional range of factors relevant to the rise of fuel poverty to be identified by researchers. This has included patterns of housing tenure, the nature of heating systems as well as socio-demographic circumstances such as household size, gender, class or education.

Scientists exploring the contingencies of energy poverty in the UK and RoI have often emphasized the deleterious health consequences of living in inadequately heated homes and the relationship between domestic energy deprivation and thermal efficiency interventions (Liddell & Morris, 2010). It has also been highlighted that energy poverty decreases the quality of life and influences social attainment. Authors working in this vein have argued that 'raising incomes can lift a household out of poverty, but rarely out of fuel poverty' (Boardman, 1991, p. xv), since residential energy inefficiency is the main reason for fuel poverty, and low-income households have to buy expensive warmth. Some researchers have pointed out that the spatial distribution of fuel poverty is highly sensitive to the way in which household incomes are measured. They have claimed that, regardless of the operational definition and measurement approach, households that need to spend more than 10 per cent of their income on energy on heating are generally not the same households as those in fact reporting difficulty in doing so.

As is argued in the sections that follow, such measurement and detection difficulties also apply to the wider European context. This chapter, therefore, first traces the evolution of work on energy poverty across Europe before reviewing the distribution and composition of energy-poor populations.

Energy Poverty in Continental Europe: Multi-sited Studies

The amount and depth of energy poverty-relevant research decrease rapidly once the focus is shifted away from the British Isles and onto continental Europe. Nevertheless, the generic causes of domestic energy deprivation in this context can be inferred from the emergent body of work pertaining to the European Union (EU) Member States and their neighbours. Similar to the UK and RoI, these arguments accepted that energy poverty in continental European countries arises out of a combination of low incomes and inefficient homes. However, it became increasingly recognized that the specific energy needs of a household—expressed via demographic circumstances such as household size, gender, occupation or class—also play a role. Of no less significance is the nature of housing tenure and heating system, since they may limit the energy efficiency interventions and fuel switching measures that can reduce energy costs (Bouzarovski & Simcock, 2017; European Commission, 2013; Pye et al., 2015).

Some of the initial non-UK and non-RoI scholarship about the energy and poverty nexus in multiple European countries included analyses of housing, fuel poverty and health in the European context, using data from the European Community Household Panel (ECHP). Contributions in this vein were based on a consensual approach, which 'unlike traditional forms of measuring relative poverty … does not rely on the opinions or scientific postulates of academics or experts' (Healy, 2017, p. xii). They combined objective housing data with 'indicators of socially perceived necessities' to demonstrate, inter alia, the central role of inefficient homes and poorly designed—or absent—heating systems in the production of energy poverty. Of note was a 14-country exploration of excess winter mortality: describing a seasonal increase in deaths that can be commonly attributed to 'cold strain from both indoors and outdoors' (Healy, 2003, p. 784). It linked information about thermal efficiency standards and mortality patterns with 'longitudinal datasets on risk factors pertaining to climate, macroeconomy, health care, lifestyle, socioeconomics, and housing' (ibid.).

The results of this investigation established that 'those countries with the poorest housing (Portugal, Greece, Ireland, the UK) demonstrate the highest excess winter mortality' (Healy, 2003, p. 788); socio-economic well-being was also shown to play a role.

Also influencing early energy poverty debates was World Health Organization-led investigation of 'housing, energy and thermal comfort' in eight European countries, plus Kazakhstan and Kyrgyzstan. Using a range of independently gathered data, many of the country case studies within this inquiry established that seasonal winter mortality was a problem across Europe. Its conclusions underlined that 'inadequate housing' is the fundamental problem in this context. The authors also argued against a pan-European definition of 'fuel poverty', emphasizing that it may be 'more appropriate to give guidance on the factors to be taken into account in developing a national definition' (World Health Organization Regional Office for Europe, 2007, p. 10).

Among the most widely cited pieces of research in this domain are the results of the aforementioned EPEE (European Fuel Poverty and Energy Efficiency) project, which used three indicators from the SILC (Statistics on Income and Living Conditions) data set ('ability to pay to keep one's home adequately warm', 'leaking roofs, damp walls/floors/foundation, or rot on window frames/floors', 'arrears on utility bills') to evaluate the extent of fuel poverty in Belgium, Spain, France, Italy and the UK. This data was then cross-referenced with information from other demographic indicators in SILC, as well as national surveys about the level of household incomes, as well as the nature of the housing stock and heating system. The study emphasized that one in seven households in Europe is in or at the margins of 'fuel poverty', locating the causes of the condition within the familiar context of low household incomes, insufficient heating and insulation standards and high energy prices.

Moving further east, the World Bank also sponsored an investigation of heating strategies among the urban poor in Croatia, Latvia, Lithuania, Moldova as well as Armenia, Kyrgyzstan and Tajikistan (Lampietti & Meyer, 2002). Even though this inquiry did not use an explicit 'energy poverty' lexicon, it did offer a broad-level investigation of household energy consumption and heating patterns in the selected countries. Having illuminated the wider relationship among heating, poverty alleviation and environmental quality issues, the research provided a series of policy suggestions about the necessary steps to design policies that will enable the provision of 'clean heat' in 'fiscally sustainable ways' (Lampietti & Meyer, 2002, p. 23). The study built on previous World Bank-led work in the region (Buckley & Gurenko, 1997; World Bank, 1999a, b).

Working along similar lines but with a stronger focus on social policy issues was an exploration of the social safety nets for energy price increases used by Bulgaria and Romania, in addition to Armenia and Kazakhstan (Velody, Cain, & Philips, 2003). Having established that 'energy costs are the highest monthly expense after food for most low-income households in the region' (Velody et al., 2003, p. vii), the study examined the poverty alleviation role played by three types of mechanisms: fuel assistance payments, energy efficiency improvements in low-income residences and 'progressive' tariff structures. It concluded that social protection instruments at the energy–poverty nexus were most effective if they provided a well-targeted and meaningful level of assistance, and were implemented via stand-alone and easily manageable mechanisms. The results of this work were echoed in a report on power sector affordability in South East Europe, which, having undertaken a series of analyses, found that many South East European countries have not yet developed adequate social safety mechanisms to protect energy-poor consumers (European Bank for Reconstruction and Development, 2003, p. 2).

There is also research that operates at a broader geographical scale, but in more narrow conceptual terms. A working paper published by the European Bank for Reconstruction and Development examined how 'energy burdens' (the share of household income devoted to energy) would change across 27 post-socialist countries in Eastern and Central Europe (ECE) and the Former Soviet Union (FSU) in a situation where 'all utility prices are raised steadily to reach full cost recovery levels by 2007' (Fankhauser & Tepic, 2007, p. 15). Having noted that 'it is surprising how little we still know about the consumption patterns and well being of low income households' (ibid.), its authors claimed that 'delaying tariff adjustments may not be an effective way of mitigating the social impact of tariff reform' (Fankhauser & Tepic, 2007, p. 15).

In their entirety, such studies confirm that one of the key driving forces of energy poverty in the Eastern European context have been energy price increases undertaken after the fall of communism, so as to bring electricity and gas tariffs—formerly subject to indirect subsidies by the state—up to cost-recovery levels. Work in ECE and FSU has thus revealed a series of 'pervasive geographies' of energy poverty arising from the failure of the state to respond to price increases with adequate social welfare support and energy efficiency investment (Buzar, 2007b). Other key driving factors include tenure patterns within the housing stock, as well as the regulation of energy markets; more recently a further complication has been added by the effects of the financial crisis and associated mortgage payments (Maxim, Mihai, Apostoaie, & Maxim, 2017). Nevertheless, a lack of unified approaches has been noted, in addition to the disproportionate coverage of energy poverty within social policy.

Much of this work has highlighted the significant difficulties faced by disadvantaged households in the region. It has demonstrated that, in addition to affordability and energy efficiency issues, important dimensions in the rise of energy poverty include the nature of household energy needs, as well as the fact that some demographic groups are 'trapped' in housing arrangements and heating systems that do not allow for switching towards less expensive and more comfortable ways of providing energy services (Buzar, 2007c; Tirado Herrero & Urge-Vorsatz, 2012). Research focusing on the relationship between a household's awareness of climate change issues, on one hand, and energy efficiency retrofits, on the other, has also provided a range of energy poverty-relevant insights (Bouzarovski, 2015). It has highlighted that the poor quality of the housing stock may combine with the feeling of being too cold, hot or uncomfortable in driving energy-related renovations in privately owned dwellings (Bartiaux et al., 2012; Cirman, Mandič, & Zorić, 2013).

Energy poverty-relevant evidence can also be found in a study of 'the effects of energy reforms on the probability of households experiencing deprivation, defined as difficulty in paying the bills' (Rezessy, Dimitrov, Ürge-Vorsatz, & Baruch, 2006, p. 253). Authors working in this vein provide a range of statistical analyses of ECHP and SILC data for Denmark, Belgium, France, Ireland, Italy, the Netherlands, Spain, Austria, Finland, Luxembourg, Norway and Sweden. Their conclusions highlight that 'unbundling vertically integrated activities in the electricity sector and reducing public ownership in the gas sector are both correlated with higher probability of experiencing deprivation' (Rezessy et al., 2006, p. 262). Academics have also used statistical analyses of SILC and European Quality of Life Survey data to explore the relationship between self-reported energy poverty-relevant indicators and other socio-demographic and spatial variables (Thomson & Snell, 2013; Thomson, Snell, & Bouzarovski, 2017).

In-Depth Research at the National and Local Scale

Research relevant to the causes and consequences of domestic energy deprivation has also been produced in relation to the circumstances of particular countries. One of the most influential debates in this regard commenced with a highly publicized paper on the welfare effects of raising household energy prices in Poland (Freund & Wallich, 1996). Its empirical analysis was based on data from the 1993 Polish household budget survey, 'which contains information on the expenditures of 16,044 Polish households, surveyed between January and June 1993' (Freund & Wallich, 1996, p. 55). Examining the expenditure patterns of households in five equivalent income quintiles led the authors to conclude that 'not only did the better off spend a larger absolute amount on energy than the poor, they also consumed a larger proportion of their expenditures as energy' (ibid.). A similar analytical approach was used in research of the extent to which 'electricity tariff increases in Ukraine hurt the poor' (Dodonov, Opitz, & Pfaffenberger, 2004, p. 855), whose authors recommended that price increases up to levels comparable to those in OECD (Organisation for Economic Cooperation and Development) countries 'should only be realized in steps' (ibid.).

The results of these studies have been favourably received in policy circles; the fact that their findings chimed in with the neoliberal agenda for energy sector unbundling and privatization pursued throughout Europe—and particularly in the East—after 1989 has allowed them to be widely cited in the literature on energy sector reform even though the use of elasticities and consumer surplus to estimate social welfare in conditions of 'very high price increases' has been problematized by some (Bacon, 1994). Many policy discussions of the distributional consequences of energy restructuring have widely cited the finding that implicit energy price subsidies benefit the 'rich' more than the 'poor' (Buzar, 2007a).

Also focused on issues of energy affordability as they relate to price increases is an investigation of the distributional effects of regulatory reforms in the Italian water and energy utility sectors (Miniaci, Scarpa, & Valbonesi, 2008). Using a range of regional, demographic and climatic indicators, its authors have constructed an affordability index for public utility consumption, so as to overcome the absence of an official energy poverty definition in Italy. Their findings, which are based on statistical modelling of large data sets from the Italian family budget survey, indicate that 'in the period considered, reforms in the water, natural gas and electricity markets were not accompanied by exacerbated affordability issues in Italy' (Miniaci et al., 2008, p. 162). More recent work has updated and developed these findings further, by exploring the wider context of network industry reform, as well as the role of regional policies (Florio, 2013; Scarpellini, Sanz Hernández, Llera-Sastresa, Aranda, & López Rodríguez, 2017).

There have also been several multinational organization-led studies using an explicit energy poverty framework in the design of research methods and approaches. A United Nations Development Programme-supported investigation in Serbia and Montenegro provided an integrated and comprehensive take on the relationship between energy, poverty and environmental problems. It introduced access considerations to the equation, by distinguishing between indicators relevant to the provision of energy services—including fuel consumption and the use of household energy appliances—and measures of the sufficiency of energy services, such as space heating, ventilation, domestic hot water and cooking (Kovačević, 2004).

Other ECE states have also been the subject of scientific attention in the field of domestic energy deprivation. The expansion of energy poverty in Bulgaria has been documented using interview and national household survey data, and with reference to EU and national policies. Part of the context for such work stems from the fact that in addition to having some of the highest rates of households reporting inadequate domestic thermal comfort in the SILC survey, this country has also implemented extensive energy privatization and liberalization reforms during the past 15 years (Bouzarovski, Petrova, & Sarlamanov, 2012; Lenz & Grgurev, 2017). Energy poverty in Poland has also been extensively studied and described thanks to the work of the Institute of Structural Research as well as several academic researchers (Miazga & Owczarek, 2015).

As we argued in a recent co-authored paper (Bouzarovski, Tirado Herrero, Petrova, & Ürge-Vorsatz, 2016), successive Hungarian governments have been making various attempts to buffer the impact of growing energy prices on the purchasing power of Hungarian households and voters. These politically motivated policy interventions have mainly taken the form of regulated energy prices and relatively short-lived subsidy schemes. Utility cuts are firmly entangled in political strategies to gain electoral support by confronting EU institutions and international corporations. Presented as the 'battle of the utility bills' (*rezsiharc*), such efforts were a central theme of the right-wing government's campaign before the April 2014 general elections, in which the populist Fidész party achieved a new parliamentary majority. While it is likely that the measures have brought short-term benefits to low-income households by allowing for a reduction in energy burdens, their ability to address the wider spatial and infrastructural components of energy poverty and vulnerability is questionable. This is because they have preferentially supported urban consumers of natural gas and district heating, while failing to provide relief to households (mostly in rural areas) relying on bottled natural gas or firewood as a source of heat (Szivós, Bernát, & Kőszeghy, 2011). Also, there have been fears that the subsidies may increase rates of energy poverty by diverting resources that could be used for reducing the country's supply dependency on Russia, or investing in residential energy efficiency.

Significant forays are also being made into scientific understandings of the underlying causes of energy poverty in various Southern European countries, where the condition has received almost no academic attention to date. This includes insights into the causes and patterns of energy poverty in Spain (Phimister, Vera-Toscano, & Roberts, 2015; Sánchez-Guevara Sánchez, Mavrogianni, & Neila González, 2017; Tirado & Jiménez Meneses, 2016); such work has demonstrated the existence of a close link between unemployment and energy poverty, in addition to establishing that existing social safety nets are failing to provide adequate assistance to energy-poor populations. An extensive study in the Greek capital Athens has uncovered the links between low incomes and energy efficiency by establishing that 'low income people are more likely to be living in old buildings with poor envelope conditions' (Santamouris et al., 2007, p. 893). Operating on a vastly different—but no less relevant—scale, research of energy-saving interventions in this country's mountainous areas has led the author to conclude that 'utilizing locally produced biomass and applying energy-saving measures can bring households below the energy poverty limit' (Katsoulakos, 2011, p. 284). The geographies of energy poverty in Greece and Cyprus are now among the most studied in Europe (Atsalis, Mirasgedis, Tourkolias, & Diakoulaki, 2016; Boemi, Avdimiotis, & Papadopoulos, 2017; Boemi & Papadopoulos, 2017; Katsoulakos & Kaliampakos, 2016; Papada & Kaliampakos, 2017; Petrova, 2017; Santamouris et al., 2014), and work by Greek researchers has also contributed to an improved understanding of indoor conditions across Europe (Kolokotsa & Santamouris, 2015).

Western European countries have also attracted significant new interest. An exploration of the everyday strategies that are employed by Austrian households in order to alleviate domestic energy deprivation has revealed that 'energy-inefficient windows, buildings and housing sites are the cause of heavy burdens' (Brunner, Spitzer, & Christanell, 2012, p. 7) for this group. Conceptualizing processes of targeting, identification of households and implementation as three interdependent steps has highlighted the complex errors of inclusion and exclusion implicated in the design of France's rapidly developing fuel poverty policy (Dubois, 2012). The increasing amount of public attention and state funding attracted by the energy poverty predicament in this country has been accompanied by the expansion of scientific research devoted to the issue, especially in terms of the relationship between vulnerability patterns and support policies (Bafoil, Fodor, & le Roux, 2014; Bartl, 2010; Legendre & Ricci, 2015; Ortar, 2016). Several recent contributions indicate that energy poverty is even present in countries like Germany, where rates of social inequality and inefficient housing are at record low levels (Becker, Kouschil, & Naumann, 2014; Billen, 2008; Großmann, Schaffrin, & Smigiel, 2016; Kopatz, 2009; März, 2017; Tews, 2014).

Geographic Patterns of Energy Poverty in Europe

I now turn to the social and spatial patterns of domestic energy deprivation across Europe, which, as evidenced by some of the work reviewed above, are highly geographically variable and locally contingent. General insights about the geographic extent of energy poverty in the EU can be gleaned from published SILC data. Based on the consensual approach (Healy, 2017) the information generated by the subjective measure on 'inability to keep the home warm' can be combined with more objective data about the shares of each country's population facing disproportionately high housing burdens, living in low-quality dwellings or having arrears on utility bills. Dividing each of these objective indicators by 3 (as they do not necessarily demonstrate energy poverty themselves), and adding them to the more direct subjective measure referring to the level of domestic heating, offers broad information about Europe's spatial patterns of energy poverty (Fig. 4.1).

The highest shares of populations with insufficient self-reported domestic warmth are concentrated in the part of the EU that is constituted by the post-socialist states of ECE (also referred to as the EU-10), especially

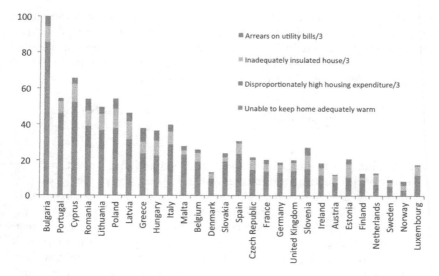

Fig. 4.1 A composite fuel poverty indicator based on the shares of populations in different EU countries facing selected energy poverty-related problems, with the values of the three 'objective' measures divided by 3. Originally published in Bouzarovski (2014)

Bulgaria. In such countries, the share of the population reporting inadequately heated homes has been 20.0 per cent, while the value of the composite fuel poverty indicator is 44.5 per cent. This is against EU-wide averages of 12.8 and 31.7 per cent, respectively. Also scoring high according to the same criteria are the eight EU countries that border the Mediterranean Sea, where 16.6 per cent of the population has reported being 'unable to keep their home adequately warm', while the composite fuel poverty indicator reaches 43.58 per cent.

As we argued in Bouzarovski and Tirado Herrero (2017b) existing knowledge thus suggests a macro-regionalization of the EU in several clusters of countries with different energy poverty levels and dynamics. In order to explore the consistency of this categorization with respect to correlation analysis presented in the previous section, we plotted the average value of Eurostat's monetary deprivation indicator 'at-risk-of-poverty' rate (percentage of the population with an income below 60 per cent of the national median, after social transfers) against an ad hoc composite energy poverty index for each member state. The energy poverty index took into account the SILC population percentages of people who have reported (i) being unable to keep their homes adequately warm (*Inability*); (ii) having arrears in utility bills (*Arrears*); and (iii) living in a home with a leaking roof, or the presence of damp and rot (*Housing faults*):

$$\text{Energy poverty index} = \begin{pmatrix} 0.5 \times \%\text{Inability} + 0.25 \times \%\text{Arrears} + \\ 0.25 \times \%\text{Housing faults} \end{pmatrix} \times 100$$

In the index, the indicator *Inability* received a higher weight in order to reflect the greater importance that our assessment gives to self-reported thermal discomfort levels in comparison with the indicator *Arrears*, which keeps track of late payment levels in energy and other utility bills. At the same time, *Housing faults* is closely related to, but not necessarily a direct indicator of, energy poverty. Our weighted values approach was thus based on previously developed energy poverty indices and weight values (Thomson & Snell, 2013). It operated under the premise that consensual measures (such as the self-reported inability to keep warm) are insufficient to capture the complex economic and material underpinnings of energy poverty, and should be combined with indicators describing the housing and financial conditions of the population in order to obtain a fuller picture.

The resulting bivariate comparison (Table 4.1) showed a low degree of positive linear correlation between the energy poverty index and the at-risk-of-poverty rate, even though relatively high levels of positive and statistically significant linear correlations were found to exist on an indicator-by-indicator

Table 4.1 Correlation matrix: Pearson's r coefficients of linear correlation between SILC energy poverty indicators and index (columns) and the at-risk-of-poverty rate (rows), calculated upon average values of EU-28 Member States for the period 2003–2013. Originally published in Bouzarovski and Tirado Herrero (2017a)

	Inability	Arrears	Housing faults	Energy poverty index
At-risk-of-poverty rate (after social transfers)	0.523**	0.574**	0.480**	0.264

**$p < 0.01$; *$p < 0.05$ level

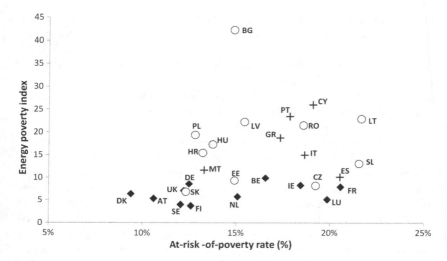

Fig. 4.2 Percentage of people at risk of poverty versus the energy poverty index. Average for EU member states 2003–2013 for both variables. Originally published in Bouzarovski and Tirado Herrero (2017b)

basis. In terms of macro-regions identified for the spatial analysis of energy poverty trends in the EU (Fig. 4.2), Western and Northern countries (noted in black diamonds) belong to a compact cluster reporting low energy poverty levels in relation to the at-risk-of-poverty rate. At the same time, Southern (crosses) and ECE Member States (circles) form a more heterogeneous group. They are characterized by energy poverty index values that are higher in relation to their at-risk-of-poverty-rates. With respect to the measurement of poverty and social exclusion, these results highlight the importance of material and housing deprivation dimensions, such as the inability to keep the home adequately warm. They emphasize the need for moving beyond purely monetary indicators, such as the at-risk-of-poverty rate.

Thus, it can be argued that a core versus periphery distribution is a better descriptor of the spatial disparities in energy poverty rates across the EU than the traditional three-region model. The resulting European infrastructural divide is enmeshed in the improved macroeconomic performance and income levels among the latter group of states, as well as their higher-quality housing stock and more effective targeting of vulnerable groups. Overall, the principal differences between core and periphery countries are reflected in the degree of public recognition received by energy poverty, its socio-demographic extent as well as the structural drivers of the condition. While cultural differences may partly explain the disproportionately high prevalence of self-reported inadequately heated homes in Eastern, Central and Southern Europe, there is little doubt that energy poverty is objectively present in these parts of the continent to a much higher extent than elsewhere.

The structural causes of energy poverty in the two regions, however, are markedly different. As was pointed out in the previous sections of this chapter, ECE states have provided fertile ground for the expansion of energy poverty due to the unique combination of cold climates, above-average rates of inefficient residential buildings, insufficiently developed and/or decaying infrastructure, high rates of income inequality and systemic issues in the management of energy, social welfare and housing operations. The socialist centrally planned economy left behind an energy sector that was entirely state-owned and -run, with indirect cross-subsidies from industry to the residential sector creating a pricing structure whereby household energy tariffs were set at below cost-recovery levels. Consequently, most countries in the region undertook dramatic price increases in order to remove such subsidies, while unbundling and privatizing energy companies so as to open up the industry to competition.

During the post-socialist transition, however, most governments were unable to provide adequate social assistance and energy efficiency investment to protect vulnerable households from energy price increases. This meant that many families had no option other than to cut back on their energy purchases. The concurrent rise in income inequality and overall poverty, alongside the initial lack of concerted efforts to improve the energy efficiency of rapidly decaying housing stocks and energy infrastructures, has created a situation whereby energy poverty now includes large parts of the population. In Poland, for example, the average 'energy burden' (the share of energy expenditure within total household expenditure) has been steadily increasing throughout the post-socialist period, even though both absolute and relative poverty have fallen during the same period. This suggests that energy affordability problems are widespread among the population, and that the expansion of economic prosperity is failing to relieve the pressure of rising energy costs on household budgets.

In Bouzarovski et al. (2016) we found that energy burdens have been on the rise particularly rapidly in Hungary: from 11.6 per cent in 2005 to

16.9 per cent in 2011. While the figures for Hungary point to the pervasive presence of energy poverty across the country, neighbouring Poland and Czechia also face difficult circumstances in this regard—in light of the fact that the literature on the subject considers energy burdens near or at 10 per cent as a sign of hardship (Boardman, 1991; Fankhauser & Tepic, 2007), it is notable that significant numbers of households in all three countries have energy burdens above 20 per cent. This is where the comparatively greater size of the problem in Poland becomes more visible, as does the significant recent increase of the population affected by domestic energy deprivation in Hungary. In more general terms, it becomes apparent that post-communist energy sector reforms undertaken in all three countries—as well as the ability of nations like Poland to maintain GDP growth after the post-2008 financial crisis—have not translated into decreased energy costs or burdens for the general population and vulnerable groups alike.

The high prevalence of energy poverty in Southern European countries has been attributed to the lack of adequate heating systems, as well as the overall poor quality of residential dwellings, which has resulted in insufficient thermal insulation. In 2004, it was reported that only 12, 8, 6 and 16 per cent of Greek households had, respectively, cavity wall insulation, double-glazing, floor insulation and roof insulation in their homes (Healy, 2017). The situation was worse in Portugal, where the corresponding figures were 6, 3, 2 and 6 per cent. Nearly a quarter of Portuguese households had stated that they had rotten window frames, while a third revealed that they had patches of condensation on the indoor walls of their home (both of these conditions are considered good indicators of poor energy efficiency). Moreover, the same study found that 16, 19 and 11 per cent of households in, respectively, Greece, Portugal and Spain are suffering from leaking roofs, indicating the absence of adequate roof insulation. An additional problem in Mediterranean states is posed by the need for cooling. According to SILC data, 30 per cent of the population in the eight states bordering the Mediterranean Sea have reported that they are unable to keep their homes adequately cool in summer. Almost two-thirds of such households are considered income poor, while 70 per cent of them are above 65 years of age.

Countries such as the RoI, the UK—and to a lesser extent Belgium and France—constitute a third geographical realm with above-average rates of energy poverty in the EU. For example, it has been reported that indoor damp, itself a very strong indicator of energy poverty, is particularly prevalent in these countries (Healy & Clinch, 2002). For the reasons outlined above, the RoI and the UK developed a wide range of measures to combat the problem: in the UK, the Warm Homes and Energy Conservation Act, effective November 2000, resulted in the implementation of an unprecedented set of policies for fuel poverty reduction, embodied in the 2001 UK Fuel Poverty Strategy. According to this document, fuel poverty reduction targets should have been achieved by eliminating fuel poverty among 'vulnerable' households (older persons, sick and disabled households and families with children) by 2010, expanding to all households by 2016.

The large-scale geographic variations discussed above mean that energy poverty is particularly concentrated in Southeastern Europe, where millions of households are likely to be suffering from a lack of adequate domestic energy services. Conventionally vulnerable groups such as 'pensioners, unemployed, low income households' have been particularly hard-hit, especially in the states that have not yet developed 'adequate social safety mechanisms' to protect energy-poor consumers. The limited extent of certain types of networked energy infrastructures (particularly gas) means that, in addition to inefficient residential stocks and affordability issues, energy deprivation is also predicated upon the spatial and technical limitations associated with switching towards more affordable fuel sources in the home. The demise of district heating systems—associated with spiralling supply costs and vicious cycles of disconnection, and coupled with rapidly rising electricity prices—has meant that some parts of the population have had no option other than using fuel wood for heating. This is particularly evident in Bulgaria, where switching towards this source of energy has a clear income dimension (Bouzarovski et al., 2012).

The substitution of modern energy carriers—mostly natural gas—by traditional or solid fuels for domestic energy heating has been reported in several ECE states (Fankhauser & Tepic, 2007; UNDP, 2004). It is evidenced by the fact that approximately 36 per cent of Hungarian households were relying on solid fuels in 2011, as opposed to 14 per cent in 2005. The trend transpired despite the presence of piped gas links in 76 per cent of dwellings and 96 per cent of settlements in Hungary, even if the amount of natural gas consumed per household dropped from 1457 m^3 per year in 2005 to 934 m^3 per year in 2011. The reliance on solid fuels has displayed a powerful income differential, with over half of all households in the bottom income decile resorting to this source of energy (Table 4.2). The propensity for lower-income households to consume solid fuels is indicative of the increasing inequality in the purchasing power of households, rather than matters of evolving cultural or economic preferences.

Table 4.2 Percentage of Hungarian households who dedicated more than 10 per cent of their energy expenditure to solid fuels in 2005 and 2011, by income deciles. Originally published in Bouzarovski et al. (2016)

Income deciles	1	2	3	4	5	6	7	8	9	10
2005	57	39	33	31	25	25	24	17	12	5
2011	60	48	48	44	43	39	33	31	21	13

As a result of these developments, firewood—the dominant solid fuel consumed by Hungarian households, alongside coal and woodchips—now trails natural gas as the second most common energy carrier for domestic space heating, even though both fuels are often used synchronously.

Difference Within Countries, Regions and Social Groups

Overall, the academic literature has found above-average rates of energy poverty among older people, families with children, and households with disabilities, long-term illness, or infirmity (Bouzarovski, 2014). In the Irish context, for instance, 'over half of elderly households endure inadequate ambient household temperatures during winter' (Healy & Clinch, 2002, p. 329). The EPEE project has also identified as vulnerable populations those out of work or in poorly paid jobs, and those dependent on social security benefits. Earlier, it was established that the group most susceptible to persistent energy poverty in the 'older' EU-15 states is single parents, followed by lone pensioners (Gray, 1995). It has also transpired that households living in multi-family apartment blocks are more likely to be suffering from energy poverty if they live in Northern as opposed to Southern Europe, partly due to income differentials. Tenure has also shown to be an important predictor of energy poverty, with households living in rental homes more vulnerable to the condition (Bouzarovski, 2014).

The scale of the energy burden is often a good predictor of the socio-demographic groups suffering from energy poverty. In Poland, for example, disproportionate expenditure on energy is correlated to household size among pensioners, with lone pensioners facing particular difficulties (ibid.). Above-average rates of energy expenditure can also be found in the case of all households headed by 'manual' workers and farmers. On average, large households are more likely to suffer from this condition compared to medium-sized households. Similar trends can be found across other Eastern European states.

Micro-scale social and residential typologies of energy poverty aggregate across broader spaces and scales to produce specific geographical patterns of vulnerability. Thus, Household Budget Survey (HBS) data for Hungary show that households with high energy burdens and facing a situation of low energy and high incomes (alike the LIHC [Low Income High Cost] indicator described in Chap. 2) are disproportionately concentrated in suburban areas, villages and areas with 'poor housing' as defined by the national statistical office (Table 4.3). This confirms previous indications about the prevalence of domestic energy deprivation in rural areas. With the exception of the 'poor housing' category, self-reported inadequate domestic

Table 4.3 Energy poverty indicators for selected housing typologies in Hungary (expressed as shares of households in the relevant category within all households). Above-average values are italicized and shaded. Originally published in Bouzarovski and Tirado Herrero (2017a)

Indicators	Energy burden exceeds 20 per cent	LIHC	Dwelling uncomfortably warm in winter	Dwelling uncomfortably cool in summer
Total household share	31	13	20	27
Urban area	21	7	*24*	*31*
Housing estate, apartment block	19	7	15	*52*
'Garden suburb'*	15	7	14	18
Suburban area**	*36*	*14*	*21*	17
Village	*41*	*19*	20	16
Industrial area	19	7	*37*	*40*
Area with poor housing	*36*	*16*	*73*	*53*
Other	*66*	13	22	22

*Cottages, dwellings in multi-apartment buildings
**Detached houses

heating and cooling rates diverge from such metrics, as evidenced by the above-average concentration of households experiencing such conditions in urban and industrial areas, as well as suburbs in the case of heating and apartment blocks in the case of cooling. Overall, this points to the influence of housing stock characteristics in influencing the quality of the final energy services received by households.

The spatial distributions of above 20 per cent energy burden and LIHC household shares in Poland and Czechia exhibit similar spatial patterns, with areas of low and medium population density hosting above-average numbers of families experiencing such difficulties (Table 4.4). Inadequately

Table 4.4 Energy poverty indicators for selected housing typologies in Hungary (expressed as shares of households in the relevant category within all households). Above-average values of the 'category' shares are italicized and shaded. Originally published in Bouzarovski and Tirado Herrero (2017a)

Indicator	Energy burden exceeds 20 per cent		LIHC		Dwelling insufficiently warm in winter		Dwelling insufficiently cool in summer	
Household share	Category	Total	Category	Total	Category	Total	Category	Total
Area with a high population density								
Poland	15	20	10	14	*14*	13	*24*	19
Czechia	8	12	6	9	-	-	-	-
Medium population density								
Poland	*21*	20	*15*	14	11	13	17	19
Czechia	*13*	12	*10*	9	-	-	-	-
Low population density								
Poland	*25*	20	*18*	14	*14*	13	16	19
Czechia	*15*	12	*11*	9	-	-	-	-

cool and warm homes in these two countries, however, are generally more present in high-density regions, possibly pointing to the urban character of such circumstances.

Energy deprivation indicators also exhibit different forms of geographical variation within the three study countries. Capital city regions in Czechia, Hungary and Poland alike are notable for the low concentrations of households with high LIHC scores or energy burdens (Fig. 4.3). The share of

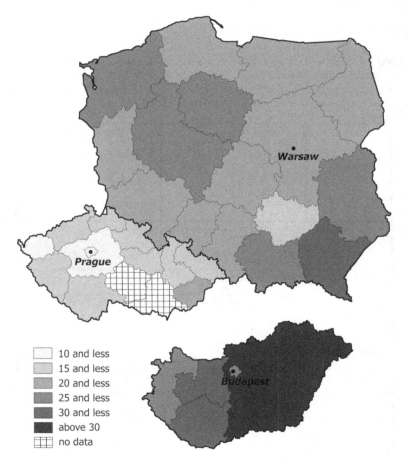

Fig. 4.3 Regional variation in the shares of households that experienced energy burdens above 20 per cent in three Central European countries. Originally published in Bouzarovski and Tirado Herrero (2017a)

households experiencing energy burdens above 20 per cent follows is a clear east–west gradient in Hungary—with the predominantly agricultural and economically underdeveloped eastern parts of the country registering record percentage levels of this indicator. A more concentric pattern (focusing on Prague) seems to be present in Czechia, with the thinly populated resource periphery of the Zlín region ranking the highest according to this statistic. The northeastern and southeastern parts of Poland are more vulnerable than the rest of the country, especially the rural Podkarpackie region at the border with Ukraine and Slovakia. The relatively privileged position of capital cities can also be seen in the regional patterns of LIHC household shares (Fig. 4.4), even if a more differentiated picture emerges at the national scale: the highest values can be found in southeastern Hungary (the northeast seems to be faring relatively better) as well as a number of regions in central, northeastern and northwestern Poland. This is despite the fact that the far southeast still has the highest percentages of households with LIHC. The Zlín region is still the most vulnerable in Czechia in LIHC terms, albeit this statistic also identifies two neighbouring Northern Moravia regions as well as the far northwest Karlovy Vary region as susceptible to the condition.

It is important to note the lack of a direct correspondence between above-average household percentages of the energy burdens and LIHC indicators, on the one hand, and per capita GDP values, on the other. Thus, the lowest levels of per capita economic output can be found in the Hungarian northeast, even if LIHC percentages are highest in the southeast (Fig. 4.5). The deprived northwest and northeast regions of Czechia do not appear to concentrate above-average numbers of energy-poor households. The discrepancy between more conventional patterns of economic inequality and domestic energy deprivation indicators is also apparent in Poland, where, for example, the relatively underdeveloped Opolskie and Lubuskie Voivodeships close to, respectively, the Czech and German borders rank relatively low on the energy burden and LIHC scores; the same applies to the entire northeast of the country, where GDP per capita levels are even lower. The picture becomes even more complex if self-reported levels of inadequate domestic heating or cooling are explored at the regional scale (Fig. 4.5). Polish regions hosting larger urban centres (e.g. Warsaw, Wroclaw Lodz and Szcecin) appear to be more vulnerable according to these measures. In Hungary, some of the highest values have been reported for the capital Budapest and its surroundings. Notably, three Eastern Hungarian regions are characterized by higher or equal percentages of households who feel their home is poorly

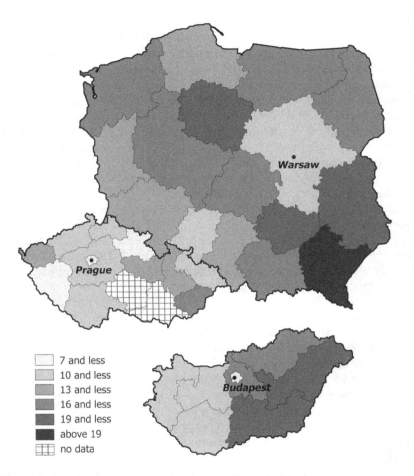

Fig. 4.4 Regional variation in the shares of households that were energy poor according to the LIHC indicator, in three Central European countries. Originally published in Bouzarovski and Tirado Herrero (2017a)

heated, when compared to being poorly cooled—a trend that diverges from the remainder of their host country, and all of Poland. When considered together with the values of expenditure-based indicators for the three regions in question, this trend may point to the severity of energy poverty in the eastern part of Hungary as a whole.

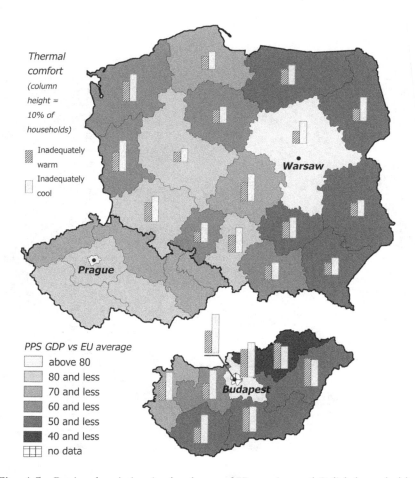

Fig. 4.5 Regional variation in the shares of Hungarian and Polish households that experienced inadequately warm or cool homes, mapped against PPP (purchasing power parity)-adjusted GDP per capita figures. Originally published in Bouzarovski and Tirado Herrero (2017a)

Czech data on the relationship between settlement size, on the one hand, and LIHC or high energy burden household shares, on the other, may help explain the broader geographical distribution of energy poverty indicators in this country (Fig. 4.6). The highest proportions of households

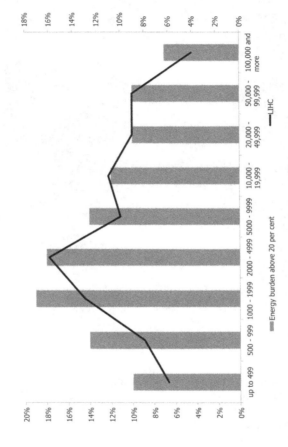

Fig. 4.6 Shares of households in the LIHC (right-hand x axis) and high energy burden (left-hand x axis) categories in different settlement size categories, Czechia. Originally published in Bouzarovski and Tirado Herrero (2017a)

with high energy burdens and costs can be found in small- to medium-sized towns. Such areas have been marginalized in housing refurbishment programmes, while facing a range of issues surrounding the legal and technical restructuring of the housing stock. The prevalence of medium- and small-sized settlements in regions like Zlín or Karlovy Vary—as opposed to the economically more deprived but also more urbanized northwestern and northeastern parts of the country—may explain the configurations of regional inequality described above. At the same time, fuel prices and availability may have played a role in the geographical distribution of energy poverty indicators in Poland and Hungary; regions where biomass and coal are more abundant perform better on the LIHC and energy burden despite the low incomes of the population. This may be due to wider price effects, as our analyses in Bouzarovski and Tirado Herrero (2017a) have also shown an increased incidence of the LIHC and high energy burden indicators in households who use solid fuels as the main source of warmth (Table 4.5). As a whole, the reviewed evidence suggests that the demand-side fuel mix and the condition of the housing stock have combined with existing patterns of deprivation to produce new spatial distributions of energy poverty at the subnational level.

Conclusion

There is little doubt that energy poverty is a pervasive problem across the EU, and is likely to expand in coming years as a result of anticipated energy price increases. For a long time, however, systematic research on issues of domestic energy deprivation in the much of the EU was scarce, especially in the countries of ECE and the Mediterranean where this condition is likely to be most pronounced. This means that, other than the UK and the RoI—which have a longer tradition of academic scholarship and policy frameworks to address the issue—energy poverty measures in many EU Member States are of an emergent nature.

Understanding the causes, content and consequences of European energy poverty is all the more pertinent in light of the increasing policy attention that is being paid to the issue within various EU institutions. In contrast to many mainstream efforts to reduce the problem to affordability or income poverty issues, however, much of the scholarship on the problem shows that the physical and institutional arrangements underlying built environment formations and everyday household practices are just as important in this context. Thus, enabling households to access energy at a materially and socially necessitated level is just as much a question of ensuring an

Table 4.5 Housing-related indicators of vulnerability to energy poverty. In the table, 'category' refers to the share of households that are considered vulnerable to the given indicator within the specific socio-demographic category; 'total' refers to the cumulative share of households in the sample (i.e. as a proportion of all households) that are considered vulnerable to the given indicator. Above-average values of the 'category' shares are italicized and shaded. Originally published in Bouzarovski and Tirado Herrero (2017a)

Indicators	Energy burden exceeds 20 per cent		LIHC		Dwelling insufficiently* warm in winter		Dwelling insufficiently* cool in summer	
Household share	Category	Total	Category	Total	Category	Total	Category	Total
Solid or liquid fuels provide the main source of indoor warmth								
Hungary	*38*	31	*16*	13	*27*	20	19	27
Poland	*25*	20	*19*	14	*35*	13	*26*	19
Czechia	*19*	12	*11*	9	-	-	-	-
Electricity provides the main source of indoor warmth								
Hungary	*39*	31	*16*	13	*34*	20	*30*	27
Poland*	*22*	20	11	14	*26*	13	*20*	19
Czechia	*15*	12	*14*	9	-	-	-	-
District heating provides the main source of indoor warmth								
Hungary	20	31	8	13	11	20	*55*	27

(*continued*)

Table 4.5 (continued)

Poland	14	20	10	14	8	13	23	19
Czechia	7	12	6	9	-	-	-	-
*Households living in dwellings constructed before 1945***								
Hungary	31	31	13	13	28	20	20	27
Poland	23	20	16	14	27	13	22	19
Czechia	17	12	10	9	-	-	-	-
Households living in dwellings constructed between 1945 and 1960								
Hungary	36	31	16	13	23	20	18	27
Poland	23	20	17	14	15	13	19	19
Czechia	14	12	13	9	-	-	-	-
*Households living in rented accommodation****								
Hungary	20	31	6	13	32	20	44	27
Poland	10	29	7	14	20	13	25	19
Czechia	7	12	6	9	-	-	-	-
*Households living in free or reduced rent accommodation*****								
Hungary	29	31	9	13	23	20	35	27
Poland	24	29	18	14	29	13	29	19
Czechia	14	12	10	9	-	-	-	-

*Also includes gas

**Does not include dwellings older than 1900 for Hungary

***In Czechia and Poland this is the non-weighted average of households living in rent-controlled and free market rental accommodation as reported by the HBS

****Free accommodation in Hungary and Czechia, reduced rent accommodation in Poland

adequate match between housing types, heating systems and household needs as it is about incomes and energy efficiency. In broader terms, therefore, building a comprehensive EU energy poverty agenda requires a conceptual shift in the mainstream theorization of domestic energy deprivation, away from the relatively narrow focus on poverty, access and energy efficiency, onto more complex and nuanced issues of household needs, built environment flexibility and social resilience.

The reviewed evidence also indicates that the driving forces of energy poverty are themselves embedded in locally specific social, political and environmental circumstances. For example, even though countries with colder climates would be expected to exhibit a greater incidence of energy poverty, the size of the population affected by domestic energy deprivation is estimated to be the lowest in Scandinavia; conversely, it has reached record levels in Southern Europe, where higher rates of income poverty and poorly insulated homes are clearly playing a determining role, in addition to the fact that many dwellings lack satisfactory heating systems. Similarly, despite possessing some of the highest energy prices in Europe, the incidence of energy poverty in Germany is judged to be significantly lower than that in, for example, Bulgaria, where energy prices are comparatively modest. In case of the latter, however, the underlying causes of the problem reside in the poor affordability of gas, electricity and heat services and the inadequate energy efficiency of the residential sector.

Much of the reviewed literature shows that energy poverty in many vulnerable 'peripheral' EU countries is expanding, while encompassing populations well beyond the low-income bracket. This is unlike better-off states—primarily in the North and West—where domestic energy deprivation seems to be predominantly concentrated among specific sociodemographic groups. Thus, the ability to capture energy poverty via the lens of income-based indicators is less meaningful in contexts where difficulties in securing adequate levels of energy services in the home are common within the general population. There is also evidence pointing to the presence of a distinct geographic distribution of energy poverty across Europe, whereby the socio-spatial underpinnings of the condition are aggregated with wider patterns of economic inequality. In many Eastern, Central and Southern EU Member States in particular, there is a tendency for domestic energy deprivation to be concentrated in rural and peripheral regions with poor-quality housing and decreased access to affordable fuels.

REFERENCES

Atsalis, A., Mirasgedis, S., Tourkolias, C., & Diakoulaki, D. (2016). Fuel poverty in Greece: Quantitative analysis and implications for policy. *Energy and Buildings, 131,* 87–98.
Bacon, R. (1994). *Measurement of welfare change caused by large price shifts.* Washington, DC: World Bank.
Bafoil, F., Fodor, F., & le Roux, D. (2014). *What energy for the Europe of the poor? A European comparison of Great Britain, France, Germany, Poland, Hungary.* Paris: Presses de Sciences Po.
Bartiaux, F., Gosselain, V., Vassileva, D., Stamova, G., Ozolina, L., & Gara, E. (2012). *Knowledge on climate change and energy saving renovations by apartment owners in Bulgaria and Latvia. A qualitative study* (Unpublished manuscript).
Bartl, M. (2010). The affordability of energy: How much protection for the vulnerable consumers? *Journal of Consumer Policy, 33,* 225–245.
Becker, S., Kouschil, K., & Naumann, M. (2014). Armut und Infrastruktur: das Beispiel Energiearmut. *Geographische Rundschau, 66,* 10–17.
Billen, G. (2008). Energie-Sozialtarife: Antwort auf drohende Energiearmut? *Wirtschaftsdienst, 88,* 489–490.
Boardman, B. (1991). *Fuel poverty: From cold homes to affordable warmth.* London: Belhaven.
Boardman, B. (2010). *Fixing fuel poverty: Challenges and solutions.* London: Routledge.
Boemi, S.-N., Avdimiotis, S., & Papadopoulos, A. M. (2017). Domestic energy deprivation in Greece: A field study. *Energy and Buildings, 144,* 167–174.
Boemi, S.-N., & Papadopoulos, A. M. (2017). Monitoring energy poverty in Northern Greece: The energy poverty phenomenon. *International Journal of Sustainable Energy, 0,* 1–15.
Bouzarovski, S. (2014). Energy poverty in the European Union: Landscapes of vulnerability. *Wiley Interdisciplinary Reviews: Energy and Environment, 3,* 276–289.
Bouzarovski, S. (2015). *Retrofitting the city: Residential flexibility, resilience and the built environment.* London: IB Tauris.
Bouzarovski, S., Petrova, S., & Sarlamanov, R. (2012). Energy poverty policies in the EU: A critical perspective. *Energy Policy, 49,* 76–82.
Bouzarovski, S., & Simcock, N. (2017). Spatializing energy justice. *Energy Policy, 107,* 640–648.
Bouzarovski, S., & Tirado Herrero, S. (2017a). Geographies of injustice: The socio-spatial determinants of energy poverty in Poland, Czechia and Hungary. *Post-Communist Economies, 29,* 27–50.

Bouzarovski, S., & Tirado Herrero, S. (2017b). The energy divide: Integrating energy transitions, regional inequalities and poverty trends in the European Union. *European Urban and Regional Studies, 24,* 69–86.

Bouzarovski, S., Tirado Herrero, S., Petrova, S., & Ürge-Vorsatz, D. (2016). Unpacking the spaces and politics of energy poverty: Path-dependencies, deprivation and fuel switching in post-communist Hungary. *Local Environment, 21,* 1151–1170.

Brunner, K.-M., Spitzer, M., & Christanell, A. (2012). Experiencing fuel poverty. Coping strategies of low-income households in Vienna/Austria. *Energy Policy, 49,* 53–59.

Buckley, R. M., & Gurenko, E. N. (1997). Housing and income distribution in Russia: Zhivago's legacy. *The World Bank Observer, 12,* 19–32.

Buzar, S. (2007a). *Energy poverty in Eastern Europe: Hidden geographies of deprivation.* Aldershot: Ashgate.

Buzar, S. (2007b). The 'hidden' geographies of energy poverty in post-socialism: Between institutions and households. *Geoforum, 38,* 224–240.

Buzar, S. (2007c). When homes become prisons: The relational spaces of post-socialist energy poverty. *Environment and Planning A, 39,* 1908–1925.

Cirman, A., Mandič, S., & Zorić, J. (2013). Decisions to renovate: Identifying key determinants in Central and Eastern European post-socialist countries. *Urban Studies, 50,* 3378–3393.

Dodonov, B., Opitz, P., & Pfaffenberger, W. (2004). How much do electricity tariff increases in Ukraine hurt the poor? *Energy Policy, 32,* 855–863.

Dubois, U. (2012). From targeting to implementation: The role of identification of fuel poor households. *Energy Policy, 49,* 107–115.

European Bank for Reconstruction and Development. (2003). *Can the poor pay for power? The affordability of electricity in South East Europe.* London: EBRD and IPA Energy.

European Commission. (2013). *Vulnerable consumer working group guidance document on vulnerable consumers, November 2013.* Brussels: European Commission, Vulnerable Consumer Working Group.

Fankhauser, S., & Tepic, S. (2007). Can poor consumers pay for energy and water? An affordability analysis for transition countries. *Energy Policy, 35,* 1038–1049.

Florio, M. (2013). *Network industries and social welfare: The experiment that reshuffled European utilities.* Oxford: Oxford University Press.

Freund, C. L., & Wallich, C. I. (1996). The welfare effects of raising household energy prices in Poland. *The Energy Journal, 17,* 53–77.

Gray, D. (1995). *Reforming the energy sector in transition economies: Selected experience and lessons.* Washington, DC: World Bank.

Großmann, K., Schaffrin, A., & Smigiel, C. (Eds.). (2016). *Energie und soziale Ungleichheit: Zur gesellschaftlichen Dimension der Energiewende in Deutschland und Europa.* Wiesbaden: Springer.

Healy, J. (2003). Excess winter mortality in Europe: A cross country analysis identifying key risk factors. *Journal of Epidemiology and Community Health, 57*, 784–789.
Healy, J. D. (2017). *Housing, fuel poverty and health: A pan-European analysis*. Abingdon/New York: Routledge.
Healy, J. D., & Clinch, J. P. (2002). Fuel poverty, thermal comfort and occupancy: Results of a national household-survey in Ireland. *Applied Energy, 73*, 329–343.
Katsoulakos, N. (2011). Combating energy poverty in mountainous areas through energy-saving interventions. *Mountain Research and Development, 31*, 284–292.
Katsoulakos, N. M., & Kaliampakos, D. C. (2016). Mountainous areas and decentralized energy planning: Insights from Greece. *Energy Policy, 91*, 174–188.
Kolokotsa, D., & Santamouris, M. (2015). Review of the indoor environmental quality and energy consumption studies for low income households in Europe. *Science of the Total Environment, 536*, 316–330.
Kopatz, M. (2009). Energiearmut in Deutschland: Brauchen wir einen Sozialtarif? *Energiewirtschaftliche Tagesfragen, 59*, 48–51.
Kovačević, A. (2004). *Stuck in the past: Energy, environment and poverty in Serbia and Montenegro*. Belgrade: United Nations Development Programme.
Lampietti, J., & Meyer, A. (2002). *When heat is a luxury: Helping the urban poor of Europe and Central Asia cope with the cold*. Washington, DC: World Bank.
Legendre, B., & Ricci, O. (2015). Measuring fuel poverty in France: Which households are the most fuel vulnerable? *Energy Economics, 49*, 620–628.
Lenz, N. V., & Grgurev, I. (2017). Assessment of energy poverty in new European Union member states: The case of Bulgaria, Croatia and Romania. *International Journal of Energy Economics and Policy, 7*, 1–8.
Liddell, C., & Morris, C. (2010). Fuel poverty and human health: A review of recent evidence. *Energy Policy, 38*, 2987–2997.
März, S. (2017). Assessing the fuel poverty vulnerability of urban neighbourhoods using a spatial multi-criteria decision analysis for the German city of Oberhausen. *Renewable and Sustainable Energy Reviews*. https://doi.org/10.1016/j.rser.2017.07.006.
Maxim, A., Mihai, C., Apostoaie, C.-M., & Maxim, A. (2017). Energy poverty in Southern and Eastern Europe: Peculiar regional issues. *European Journal of Sustainable Development, 6*, 247.
Miazga, A., & Owczarek, D. (2015). *It's cold inside – Energy poverty in Poland*. Warsaw: Institute for Structural Research.
Miniaci, R., Scarpa, C., & Valbonesi, P. (2008). Distributional effects of price reforms in the Italian utility markets. *Fiscal Studies, 29*, 135–163.
Ortar, N. (2016). Dealing with energy crises: Working and living arrangements in peri-urban France. *Transport Policy*. https://doi.org/10.1016/j.tranpol.2016.09.008.

Papada, L., & Kaliampakos, D. (2017). Energy poverty in Greek mountainous areas: A comparative study. *Journal of Mountain Science, 14*, 1229–1240.
Petrova, S. (2017). Illuminating austerity: Lighting poverty as an agent and signifier of the Greek crisis. *European Urban and Regional Studies.* https://doi.org/10.1177/0969776417720250.
Phimister, E., Vera-Toscano, E., & Roberts, D. (2015). The dynamics of energy poverty: Evidence from Spain. *Economics of Energy and Environmental Policy, 4*, 153–166.
Pye, S., Baffert, C., Brajković, J., Grgurev, I., Miglio, D. R., & Deane, P. (2015). *Energy poverty and vulnerable consumers in the energy sector across the EU: Analysis of policies and measures.* London: Insight_E.
Rezessy, S., Dimitrov, K., Ürge-Vorsatz, D., & Baruch, S. (2006). Municipalities and energy efficiency in countries in transition. Review of factors that determine municipal involvement in the markets for energy services and energy efficient equipment, or how to augment the role of municipalities as market players. *Energy Policy, 34*, 223–237.
Sánchez-Guevara Sánchez, C., Mavrogianni, A., & Neila González, F. J. (2017). On the minimal thermal habitability conditions in low income dwellings in Spain for a new definition of fuel poverty. *Building and Environment, 114*, 344–356.
Santamouris, M., Alevizos, S. M., Aslanoglou, L., Mantzios, D., Milonas, P., Sarelli, I., ... Paravantis, J. A. (2014). Freezing the poor—Indoor environmental quality in low and very low income households during the winter period in Athens. *Energy and Buildings, 70*, 61–70.
Santamouris, M., Kapsis, K., Korres, D., Livada, I., Pavlou, C., & Assimakopoulos, M. N. (2007). On the relation between the energy and social characteristics of the residential sector. *Energy and Buildings, 39*, 893–905.
Scarpellini, S., Sanz Hernández, M. A., Llera-Sastresa, E., Aranda, J. A., & López Rodríguez, M. E. (2017). The mediating role of social workers in the implementation of regional policies targeting energy poverty. *Energy Policy, 106*, 367–375.
Szivós, P., Bernát, A., & Kőszeghy, L. (2011). *Managing household debt: Hungarian country report.* Budapest, Hungary: Tárki Social Research Institute.
Tews, K. (2014). Fuel poverty in Germany: From a buzzword to a definition. *GAIA – Ecological Perspectives for Science and Society, 23*, 14–18.
Thomson, H., & Snell, C. (2013). Quantifying the prevalence of fuel poverty across the European Union. *Energy Policy, 52*, 563–572.
Thomson, H., Snell, C., & Bouzarovski, S. (2017). Health, well-being and energy poverty in Europe: A comparative study of 32 European countries. *International Journal of Environmental Research and Public Health, 14*, 584.
Tirado Herrero, S., & Urge-Vorsatz, D. (2012). Trapped in the heat: A post-communist type of fuel poverty. *Energy Policy, 49*, 60–68.

Tirado, S., & Jiménez Meneses, L. (2016). Energy poverty, crisis and austerity in Spain. *People, Place and Policy, 10*, 42–56.

UNDP. (2004). *Stuck in the past. Energy, environment and poverty in Serbia and Montenegro.* Belgrade, Serbia/Montenegro: United Nations Development Programme/Country Office Serbia and Montenegro.

Velody, M., Cain, M. J. G., & Philips, M. (2003). *A regional review of social safety net approaches in support of energy sector reform.* Washington, DC: US Agency for International Development.

World Bank. (1999a). *Non-payment in the electricity sector in Eastern Europe and the former Soviet Union.* Washington, DC: World Bank.

World Bank. (1999b). *Privatization of the power and natural gas industries in Hungary and Kazakhstan.* Washington, DC: World Bank.

World Health Organization Regional Office for Europe. (2007). *Housing, energy and thermal comfort. A review of 10 countries within the WHO European region.* Copenhagen: WHO Regional Office for Europe.

Open Access This chapter is distributed under the terms of the Creative Commons Attribution 4.0 International License (http://creativecommons.org/licenses/by/4.0/), which permits use, duplication, adaptation, distribution and reproduction in any medium or format, as long as you give appropriate credit to the original author(s) and the source, a link is provided to the Creative Commons license and any changes made are indicated.

The images or other third party material in this chapter are included in the work's Creative Commons license, unless indicated otherwise in the credit line; if such material is not included in the work's Creative Commons license and the respective action is not permitted by statutory regulation, users will need to obtain permission from the license holder to duplicate, adapt or reproduce the material.

CHAPTER 5

Concluding Thoughts: Embracing and Capturing Complexity

Abstract This chapter summarizes the key findings of the book in the context of its purpose and frameworks, while recommending possible avenues for future research and policy.

Keywords Energy poverty • Energy vulnerability • Energy justice • European Union

INTRODUCTION

This book has explored the multiple layers of systemic change implicated in the emergence of energy poverty in Europe and beyond. I have sought to highlight the embeddedness of vulnerability to domestic energy deprivation in wider dynamics of organizational and social transformation. Of particular importance to this argument has been the need to understand how energy poverty and vulnerability are both rooted in and arise out of the material and technical features that characterize the existence of an infrastructural divide in Europe. The divide operates at multiple levels and scales of activity—from the differences between nation states, to variations within the fabric of neighbourhoods and even households themselves. The divide itself can be seen as a socio-technical assemblage that is continuously dismantled and put together by multiple political interests and path dependencies. It is highly territorially contingent, which means that the

© The Author(s) 2018
S. Bouzarovski, *Energy Poverty*,
https://doi.org/10.1007/978-3-319-69299-9_5

geographical characteristics of cities, regions and countries themselves combine to produce and sustain this particular form of injustice.

As we argued in Bouzarovski et al. (2017) the emergence of energy vulnerability of a distinct spatial formation involves the interplay between concurrent processes of social change on the one hand, and the tangible and intangible features of particular places, on the other. At the same time, the dynamics that allow energy poverty to arise and persist within specific material sites also shape wider political and social processes—as well as processes of institutional change in the energy sector itself—via an additional feedback loop. Thus, the driving forces of energy poverty and vulnerability in socio-spatial terms are multidirectional and multiscalar. They involve different temporal and spatial horizons, while demonstrating the ability to shape broader political dynamics.

The reviewed evidence shows that ECE countries are characterized by record levels of energy poverty in the European context. Here, it is clear that the decision to move towards a market-based regulation of the energy sector—involving, inter alia, the liberalization of energy trade, the rebalancing of energy prices, the unbundling and privatization of energy utilities and the creation of new institutions to facilitate competition—was a crucial component of the institutional driving forces of energy-related injustices. Even if it has been argued that 'the formal remodelling of the institutional landscape has now been largely completed in many former communist countries' (Sýkora & Bouzarovski, 2012, p. 53) a tendency to reverse the movement towards market-based policies has been observed in a number of countries. This may further increase the risks that vulnerable groups face, by denying them the potential benefits of liberalized energy markets despite removing the universal support and subsidy mechanisms that characterized the centrally planned economy. Modifications of neoliberal policies are even more concentrated at the level of organizational and social practice, where corruption, clientelism and price regulation policies have contributed to the rise of a series of hybrid regulatory outcomes.

The urban scale provides a material site for amalgamating the multiple dynamics of change described within the first and second layers of transition into specific spatial formations. Energy vulnerability is imprinted in the urban landscape through existing and new forms of socio-economic segregation, access to infrastructural services and variations in built environment structures. The fact that such configurations extend beyond areas that would be typically considered low income once again points to the cross-sectoral nature of energy vulnerability, as well as its deep connections with urban processes that evolve and develop over long periods of time. This shows how energy transitions create displacements that are reflected within multiple spatio-temporal scales and thematic areas of activity. Vulnerability to domestic energy deprivation thus cannot be considered as a household issue, but rather a phenomenon that is distributed throughout the 'energy chain' (Chapman, 1989)—an issue that warrants further research in the domain of energy geographies (Calvert, 2015). As argued previously (Bouzarovski et al., 2017) such findings call for a rethink of the conceptual assumptions that inform wider sustainability transitions frameworks, by considering the material and infrastructural characteristics of place and space as contingencies that deserve customized conceptual attention.

Domestic energy deprivation does not bring about a passive and reactive set of behaviours and practices within households and institutions (Bouzarovski, Tirado Herrero, Petrova, & Ürge-Vorsatz, 2016). Rather, the diverse strategies that are articulated with respect to the condition have far-reaching effects on the systemic conditions that underpin the emergence of energy poverty. They can thus potentially challenge the triad of distribution, procedure and recognition that dominates current understandings of the injustices that underpin fuel and energy poverty (Walker & Day, 2012) by introducing notions of spatial justice into the debate (Bouzarovski & Simcock, 2017).

The work reviewed in this book also points to the need for developing a more explicit conceptual and policy link between domestic energy deprivation and the implementation of climate policies. Ex ante studies focusing on the co-benefits and multiple benefits of energy efficiency interventions (Ürge-Vorsatz, Tirado Herrero, Labzina, & Foley, 2012) have highlighted the significant welfare-enhancing effects of thermal retrofits—a key infrastructural solution often prescribed in the policy-oriented literature. However, high or increasing levels of domestic energy deprivation complicate the application of policies that promote energy vulnerability-enhancing measures, such as renewable feed-in tariffs or surcharges paid by domestic energy users irrespective of income, needs or living conditions. In a number of European countries, the expansion of energy poverty among the general population has been accompanied by the adoption of household strategies orientated towards carbon-intensive and polluting fuels, such as coal or firewood (Bouzarovski et al., 2016).

Throughout the book, I have argued in favour of developing a deeper understanding of the manner in which material deprivation both arises out of and affects the consumption of energy services within the home. This would necessitate a more nuanced theorization of the institutional and spatial contexts that shape energy-related demographic and residential vulnerabilities. The manner in which restructuring processes in the energy and housing sectors have interacted at the regional and local scales deserves particular attention in this context. Also of importance is the nature of policy recognition afforded to groups that are susceptible to the condition but remain outside the focus of present policy measures, due to the state's failure to detect the specific age, gender and locational profiles of energy-poor households. Accepting that energy poverty cannot be addressed via standard income- or economic development-based approaches, a more comprehensive conceptualization of the condition can potentially lead to the development of improved detection and measurement frameworks. The benefits of such an effort could extend beyond Europe to other parts of the Global North, where the relationship between rising energy prices and poverty levels may become a more pressing political and economic issue in the future.

On the practical side, there are significant opportunities to address the issue via demand-side energy efficiency policies—mainly in the form of deep building retrofits and appliance market transformations. Such measures are clear win-win solutions in the case of energy poverty, as they can also assist the broader process of poverty alleviation. Given the major social and geographical differences in the incidence of energy poverty within the European Union (EU), these policies are best delivered at the regional scale. A key challenge, however, lies in exposing and treating energy poverty and energy vulnerability though a political lens (Healy & Barry, 2017): seeing them as injustices that have arisen and are allowed to persist due to the presence of particular power interests and ideologies. As such, they are within the reach of the possible with regard to citizen action and wider institutional structures.

References

Bouzarovski, S., Herrero, S. T., Petrova, S., Frankowski, J., Matoušek, R., & Maltby, T. (2017). Multiple transformations: Theorizing energy vulnerability as a socio-spatial phenomenon. *Geografiska Annaler: Series B, Human Geography, 99*, 20–41.

Bouzarovski, S., & Simcock, N. (2017). Spatializing energy justice. *Energy Policy, 107*, 640–648.

Bouzarovski, S., Tirado Herrero, S., Petrova, S., & Ürge-Vorsatz, D. (2016). Unpacking the spaces and politics of energy poverty: Path-dependencies, deprivation and fuel switching in post-communist Hungary. *Local Environment, 21*, 1151–1170.

Calvert, K. (2015). From 'energy geography' to 'energy geographies' Perspectives on a fertile academic borderland. *Progress in Human Geography, 40*, 105–125.

Chapman, J. D. (1989). *Geography and energy: Commercial energy systems and national policy*. Harlow: Longman.

Healy, N., & Barry, J. (2017). Politicizing energy justice and energy system transitions: Fossil fuel divestment and a 'just transition'. *Energy Policy, 108*, 451–459.

Sýkora, L., & Bouzarovski, S. (2012). Multiple transformations: Conceptualising the post-communist urban transition. *Urban Studies, 49*, 43–60.

Ürge-Vorsatz, D., Tirado Herrero, S., Labzina, E., & Foley, P. (2012). *Employment impacts of a large-scale deep building energy retrofit programme in Poland* (Prepared for the European Climate Foundation by The Center for Climate Change and Sustainable Energy Policy (3CSEP)). Budapest: Central European University (CEU).

Walker, G., & Day, R. (2012). Fuel poverty as injustice: Integrating distribution, recognition and procedure in the struggle for affordable warmth. *Energy Policy, 49*, 69–75.

Open Access This chapter is distributed under the terms of the Creative Commons Attribution 4.0 International License (http://creativecommons.org/licenses/by/4.0/), which permits use, duplication, adaptation, distribution and reproduction in any medium or format, as long as you give appropriate credit to the original author(s) and the source, a link is provided to the Creative Commons license and any changes made are indicated.

The images or other third party material in this chapter are included in the work's Creative Commons license, unless indicated otherwise in the credit line; if such material is not included in the work's Creative Commons license and the respective action is not permitted by statutory regulation, users will need to obtain permission from the license holder to duplicate, adapt or reproduce the material.

INDEX

A
Assemblages, 5, 16, 25, 27, 109
Austria, 80, 84

B
Belgium, 50, 78, 80, 89
Bulgaria, 79, 82, 85, 90, 102

C
Citizens' Energy Forum, 62, 65
Clean Energy for all Europeans package, 2, 57
Climate change policy, 55, 60, 63, 68, 111
Council of Europe, 50
Czechia, 89, 93, 97

E
Eastern and Central Europe (ECE), 79, 82, 88, 99, 110
Energy efficiency, 4, 10, 19, 28, 42, 59, 76, 77, 79, 80, 83, 88, 89, 102, 111, 112
EU directives, 43, 54, 60
policy, 46, 53, 54, 56
Energy geographies, 5, 111
Energy justice, 26–28, 110–112
Energy poverty
 definition, 1, 10, 43, 51, 59, 67, 78, 81
 and gender, 11, 76, 77, 112
 Global South, 2, 11, 12
 health, 57, 68
 and housing, 20, 22, 29, 50, 54, 56, 68, 86
 measurement and indicators, 14, 15, 21, 58
Energy Roadmap 2050, 46, 56
Energy services, 5, 11, 13, 15–17, 76, 90
Energy transitions, 3, 5, 12, 23, 24, 46, 61, 110, 111
Energy Union, 2, 46, 59–61
Energy vulnerability, 5, 17–22, 24, 110
 vulnerable consumers, 47, 51–57, 63
 Vulnerable Consumers Working Group, 62

© The Author(s) 2018
S. Bouzarovski, *Energy Poverty*,
https://doi.org/10.1007/978-3-319-69299-9

INDEX

Energy Vulnerability and Urban Transitions in Europe project, 3
Europe 20-20-20 strategy, 46
European Commission, 46, 49, 56, 59, 61, 63, 65
European Community Household Panel, 14, 77
European Economic and Social Committee, 56, 65
European Energy Poverty Observatory, 2
European Fuel Poverty and Energy Efficiency project, 49, 78, 91
European Parliament, 47, 52, 56, 59, 65

F
First Energy Package, 44
France, 50, 78, 80, 84, 89
Fuel poverty, 10–12, 19, 55, 78

G
Germany, 84, 102
Greece, 28, 78, 83, 89

H
Hungary, 82, 88, 90, 92, 99

I
Infrastructural divide, 87, 109
Infrastructure and systems of provision, 4, 11, 17, 19, 22, 23, 28
Institutional change, 4, 17, 23, 61, 99, 110
International Energy Agency, 55
Italy, 50, 78, 80, 81

L
Low Income High Cost, 10, 92

N
North America, 2

P
Path-dependency, 5, 109
Poland, 81, 82, 88, 93, 99
Portugal, 78, 89

R
Republic of Ireland, 2, 28, 76, 80, 89, 99
Right to energy, 3, 48

S
Second Energy Package, 44
Single Market, 44
    Internal Energy Market, 52, 61
Southeastern Europe, 90
Southern Europe, 83, 99, 102
Spain, 50, 58, 78, 80, 83, 89
Statistics on Income and Living Conditions, 14, 78, 80, 84, 86
Sustainable Energy for All, 2

T
Third Energy Package, 43, 44, 50, 64
Treaty on the Functioning of the EU, 45

U
United Kingdom, 2, 10, 50, 76, 78, 89, 99

V
Vulnerable consumers
 Vulnerable Consumers Working
  Group, 65

W
Western Europe, 84, 87
World Health Organization, 10

CPSIA information can be obtained
at www.ICGtesting.com
Printed in the USA
LVHW081933090619
620633LV00012B/332/P